Pruning and Training
Systems for Modern
Olive Growing

CSIRO
PUBLISHING

Pruning and Training Systems for Modern
Olive Growing

Riccardo Gucci

Dipartimento di Coltivazione e Difesa
delle Specie Legnose, Università di
Pisa, Italy

Claudio Cantini

Istituto sulla Propagazione delle Specie
Legnose, Consiglio Nazionale delle
Ricerche, Scandicci (FI), Italy

National Library of Australia Cataloguing-in-Publication entry

Gucci, Riccardo, 1959–
Pruning and training systems for modern olive growing.

Bibliography.
Includes index.
ISBN 0 643 06443 5.

1. Olive – Australia. 2. Olive – Pruning. 3. Olive – Training
I. Cantini, Claudio. II. Title.

634.630994

Available from
CSIRO Publishing
36 Gardiner Road, Clayton VIC 3168
Private Bag 10, Clayton South VIC 3169
Australia

Telephone: +61 3 9545 8400
Email: publishing.sales@csiro.au
Website: www.publish.csiro.au

Cover photo:
The trunk of an ancient olive tree growing at Scarlino, Italy

Set in Stempel Schneidler and Frutiger
Printed by Ingram Lightning Source

Contents

Preface

The pruning of olive trees requires an understanding of plant biology and the plant's interaction with its environment. Horticulturalists and physiologists are well aware of the variability in plant responses to climatic factors or cultural practices. Conditions may vary from year to year and from site to site even within the same orchard. To match these changing conditions the pruning of trees may require adjustments in methods and refinements in techniques that can never be entirely described and predicted in a book. We hope that it will be apparent, reading through these pages, that there are several options for pruning available to the olive grower. This is the main reason why we have not limited our work to the mere description of operations or training systems. Besides diagrams and illustrations, we have included tables of data and figures derived from research to document how we came to our conclusions. Access to the experimental data should help the reader avoid common misconceptions and dogmas concerning the pruning of olive trees. Despite our effort, we are well aware of the difficulties in applying written information to field work without proper training, and we strongly recommend beginners to be careful and humble in their approach to pruning. Pruning requires experience, and no matter how simple it may look at first, the inexperienced pruner should always begin work under supervision. Only after sufficient learning can the newly acquired knowledge be applied.

We did not want to write a book for horticultural specialists only. Olive growing is now spreading to new areas where olericulturalists are obviously inexperienced. Every effort has been made to deliver the information in an accessible way. For this purpose, the first two chapters are dedicated to basic physiological and horticultural concepts. Plant physiologists and pomologists will find these chapters superficial, but we have tried to explain physiological processes at a practical level, giving no more information than is strictly needed for those without an academic background. The format, illustrations, figures and text are organized to make this book easy to read and use. In some instances we were forced to use technical terms translated from Italian, Spanish or French, because there was no equivalent word in English.

The boxed sections are included to provide a readily accessible source for the background material necessary for the reader to understand the content of each chapter. These paragraphs are not organized in a rigid fashion and are not meant to be exhaustive. We purposely have not included the description of old-fashioned, traditional training systems, or systems that are no longer economically viable and have been described extensively in the literature (Barranco et al., 1997; Ferguson et al., 1994; International Olive Oil Council, 1996; Loussert and Brousse, 1978; Morettini, 1972; Pastor, 1989).

Olive growing is now evolving after decades of stagnation and it is rapidly expanding to English-speaking countries, where this crop had never been economically important. The type of pruning and the training system are two of the major factors in successful tree performance and orchard profitability. Despite the importance of pruning in the management of olive orchards and the fact that pruning olive plants differs considerably from pruning other fruit trees, hardly any written information is available in English. Moreover, most of the literature on pruning olive trees is quite old and no existing book specifically addresses the needs of "modern olive growing", which has at least two meanings in this context. First, it refers to cultivation aimed at achieving high yields, high quality of production, and low costs. Mechanization and reduced input of labour for pruning and harvesting are the key issues in keeping costs low. Second, it implies managing the orchard in an environmentally friendly and economically viable way, with resources optimized in the long run.

There are several different ways to grow olive trees profitably. Our goal is to provide the basic knowledge to understand plant performance with different pruning strategies and training systems so that the grower will be able to select the most economically convenient method and avoid errors in pruning that may result in yield losses and/or high cultivation costs. The objective of this book is to summarise the most up-to-date information available on current pruning techniques and training systems for olive growing. The fundamental idea underlying our effort is to teach the reader, whether experienced horticulturalist or beginner, to understand the basic concepts from which will develop his/her own skills and pruning strategy.

Riccardo Gucci   30 June 1999

Claudio Cantini

Acknowledgements

We are indebted to Frank Dennis for the thorough review of most of the manuscript, to Rolando Guerriero, Balilla Sillari, Hava Rapoport and Roberto Tognetti for their valuable comments on specific chapters, and to Colin Legg and Amy Iezzoni for English revisions of Chapters 1 and 10. We are grateful to Marco Toma and Natale Bazzanti for providing the raw data from ARSIA trials, Giorgio Paiardini for making available some photographs, and Eugenio Gucci for translation from French. We also thank Piero Puntoni for drawing Figs 1.5 and 4.6, Antonella Tedesco for technical assistance, the staff of the Accademia dei Georgofili and of the Biblioteca Nazionale in Florence for their kind assistance. A special thank-you to the editor and staff of CSIRO Publishing for the many suggestions and excellent editorial work that greatly contributed to the final outcome of this work. Finally, we would like to thank colleagues and students who provided input and support during the preparation of the manuscript.

Chapter 1

Introduction

The importance of pruning olive trees has been recognized since ancient times. Both Greek and Latin authors reported on various practices for olive growing, including pruning. The Latin agronomist Columela recommended that "olive trees not be cut for eight years after planting" and underlined the usefulness of pruning thereafter because "pruning compels the olive tree to bear fruit".

In more recent ages, early descriptions of pruning olive trees date back to the sixteenth century. In the second part of the sixteenth century GiovanVittorio Soderini (1817) wrote that "the olive tree does not like to be pruned much", while Bernardo Davanzati[1] in a short chapter on olive growing in his work *Tuscan Cultivation* (1807) pointed out that "the olive trees of the 'Correggiolo' and 'Frantoio' types do not like to be touched or only want to be pruned little, but those of the 'Moraiolo' type need to be pruned yearly". Davanzati also indicated that "young plants should not be pruned for two or three years, and pruned thereafter to leave only three or four shoots tied to one or more posts". The first specific work on olive growing was written by Pietro Vettori (1762) in 1569. Vettori summarized the information from Greek and Latin authors and expressed similar concepts as Davanzati on the necessity to adapt pruning to the different types of olive trees.

At the end of the eighteenth century Giovambattista Landeschi, a priest-agronomist living in San Miniato, dedicated a few pages of his *Essays on Agriculture* to the pruning of olive trees (Landeschi, 1775). In 1782 Giovanni Battarra wrote that "olive trees should be pruned every year and even twice a year" but admitted that "some growers prune every year, some never prune". Battarra (1782) suggested eliminating all the dead or diseased wood, to put off large cuts until after the harvest in the autumn, and to prune trees to form a basket empty inside with branches expanding horizontally. In his view the pruning of young plants was limited to "tying the trees to the trainer without any cut for the first three years after planting. However, if the laterals were prevailing over the central stem they should be inclined and tied down to diminish their vigour".

In 1794 Giovanni Presta wrote a book covering most aspects of olive growing and oil production, including detailed recommendations on pruning techniques. Although Presta (1794) reported mainly about cultivation and pruning practices in the Kingdom of Naples, he was well aware of the work previously published in Tuscany. For instance, Presta correctly underlined the risks of pruning olive trees in winter, as suggested by Davanzati (1807), because of the high probability of damage by frosts in the relatively cold climate of Tuscany.

In 1801 Jean Charles Simonde reported that olive trees in the area of Pescia (Tuscany) were not very productive because they were planted so close together that they formed a wood rather than a grove. In these conditions trees were tall and slender because of mutual shading. He pointed out that peasants only removed the dead wood necessary for cooking or heating rather than pruning trees properly to stimulate new shoot growth (Sismondi, 1995).

In a lecture to the Academy of the Georgofili in Florence, Girolamo Bardi (1802) recommended adopting the "continuous small pruning", whereby few cuts were made regularly every year to avoid major cuts later on. He also remarked that "some growers cut every year, some never cut, some every other year, others every four or eight years". At that time the canopy of the olive tree was usually trained to three main shapes (basket, inverted empty cone, or cylinder), and Bardi (1802) proposed that olive trees be pruned according to their natural shape, "a compact paraboloid with a round or conical top so that most of its parts are exposed to light and air". He underlined that the contour of the paraboloid had to be adjusted according to soil fertility and the availability of water and recommended compact shapes in arid or difficult conditions.

Even then, there were different views on how to prune olive trees. Trees were less intensively pruned in the hills around Pisa than in the Florentine hills, and Antonio Bicchi, in his notes on Landeschi's *Essays on Agriculture*, pointed out that the Pisan method of pruning produced "healthier, more fruitful and beautiful plants than those pruned in the Florentine countryside". The Pisan method involved minimum pruning, whereby only the suckers were removed and trees were pruned every other year. In most cases not even the dead wood was eliminated. Another difference between the Pisan and the Florentine olive plantations lay in the cultivation system: specialized groves were more common in the surroundings of

Figure 1.1 Mixed cropping system where rows of olive trees trained to a polyconic vase are alternated with herbaceous crops and rows of grapevine. The distance between olive rows is about 15 m.

Pisa, whereas mixed cropping systems were more frequent in other parts of Tuscany. In the latter case, grapevines, olive and fruit trees were often planted in a row on top of an embankment made to reduce the slope of the hillside. Thus, tree roots played an important role in maintaining the stability of the embankment and absorbing water that would otherwise run off and cause soil erosion. In mixed cropping systems, olive trees needed to be pruned more severely to avoid shading the grapevines and the herbaceous crops (mainly cereals) growing underneath (Fig. 1.1). The care that peasants living in the Pisan mountains devoted to olive trees was also appreciated by Sismondi (1995), who reported that "olive trees were trained to large regular pyramids with foliage exposed to light, and air circulation within the canopy".

However, not until the early nineteenth century was a comprehensive book on olive growing and oil production published. Prior to that the information was often misleading or inconsistent. The work was commissioned by the Academy of the Georgofili to Giuseppe Tavanti, who dedicated more than 20 pages to the pruning of olive trees. Not only did he make a comprehensive review of the previous literature, but he also thoroughly described the biology and cultivation of this species. As for pruning, Tavanti (1819) wrote about both theoretical and practical aspects, including techniques, timing, periodicity, and the influence of factors such as soil, climate, and cultivar. He emphasized the importance of "light on sexual reproduction and oil accumulation", of "heat on vegetative vigour", and of visual assessment and experience for the correct performance of pruning. His book is a milestone in the olive literature because of the numerous concepts and practical tips reported in a well-organized format. Nevertheless, Tavanti made little mention of training systems.

At the beginning of the twentieth century the vase was the most commonly used training system. In France the best training system was considered a sort of inverted vase with three major almost upright branches, obtained by removing the lowermost branches that were longer than the leader of each branch. This vase system, also called the goblet, was recommended both for new plantings and for rejuvenating old trees. In Arizona, Crider (1922) suggested that branches of the vase be reduced to three or four in number in the first winter after planting. The branches had to be distributed evenly to form a well-balanced top, with a spacing along the trunk not closer than four to six inches. Crider (1922) also prescribed pruning at least three times a year, once in winter and twice in summer.

Despite the development and the detailed descriptions of new vase systems, the pruning of olive trees was still neglected in many regions. At the French National Congress of Oliviculture held in Marseille in 1912, it was reported that pruning was poorly done in France, Algeria and Tunisia (with the exception of the Sfax region). In many cases trees were not pruned and fruits were often impossible to pick because of excessive tree height; in other instances pruning was too severe, cuts were poorly made, and the pruned wood was left beneath the trees. Jean Sordina recommended that pruning be done to enable light and air to penetrate

Figure 1.2 Old olive tree whose canopy has been reformed by pruning. Note the difference in size between the old trunk and the new branches.

inside the canopy, that dead wood be removed and that branches be shortened to allow air circulation among trees (Societe National d'Oleiculture de France, 1913).

In Italy in the 1930s, most attention was focused on increasing yield. Pruning to reconstitute the canopy of old trees was encouraged (Fig. 1.2), while removing old trees and replanting new ones was considered a disaster (Francolini, 1935). At the same time further refinements of the vase system led to the development of the polyconic vase. This system was developed mainly by Alfredo Roventini and Secondo Tonini, two olive specialists, who introduced physiological concepts in the practice of pruning (Roventini, 1936; Tonini, 1937). Their work started from the observation that old trees were rejuvenated in an irrational way. The rejuvenation pruning consisted in either brutally chopping the plant at a certain height on the trunk or trimming the canopy to geometrical shapes.

Trees trained to a polyconic vase had a single trunk of variable height (0.8–1.6 m) and three to six major branches (typically four) orientated in different directions for maximum use of space (Fig. 1.3). Each branch was trained to a conical shape and periodically shortened to prevent the fruiting surface from moving too far from the tree centre. Particular care was devoted to the thinning of shoots inserted near the terminal end of each branch to favour the growth of the leader (Fig. 1.4). The shoots in the centre of the canopy were also thinned to allow the even distribution of light inside the canopy. The development of the polyconic vase included a considerable theoretical effort. The canopy volume was described as the result of the rotation of a rectangular triangle around its right angle (point of dichotomy in Fig. 1.5). A well-formed polyconic vase had an estimated 9 m^2 of illuminated leaf surface per cubic metre of branch volume (Roventini, 1936). The trees were planted far apart and topping cuts were necessary to restrict canopy height. The pruning required the use of ladders and could take up to three hours per tree. The polyconic vase rapidly gained favour in Italy and programmes to train skilled labour were organized and promoted. This training system well suited the physiological features of the olive tree, and, since labour was inexpensive, the cost of the required pruning was still acceptable. The polyconic vase is still widespread in central Italy, but it is no longer recommended for new plantings because of the high labour requirement for pruning and harvesting.

Figure 1.3 Mature olive tree trained to polyconic vase in traditional dryland mixed cultivation. Note the conical shape and the arrow of each branch. The tree is approximately 7 m high (courtesy of B. Sillari).

Figure 1.4 Pruning the apical part of the branch of polyconic vase. Note the thinning of lateral shoots near the distal part of the branch to favour growth of the leader. Compare the pruned branch on the left, with that still unpruned on the right (courtesy of G. Paiardini).

The traditional picture of olive growing started changing after World War II. In Italy the socio-economic structure of agriculture was upset. The old system of partnership between the landowner and the peasant that had existed in many areas was abolished and large numbers of peasants moved to cities and urban centres to work in factories and service industries. This had strong repercussions on cultivation techniques for horticultural crops, including olive. The old mixed cropping systems were abandoned and only a few examples remain as relics today. Cultural practices had to be adapted to account for the scarcity and high cost of labour. Mechanization was introduced and new pruning methods and training systems were developed to reduce the need for labour. New technologies became available in olive propagation, orchard management and oil production. Not all innovations were successful. In the 1950s, training systems that had proven valid for other fruit crops, like the palmette, were introduced in olive growing as well, but after initial interest they were abandoned. The idea behind the development of the palmette was to increase production by increasing the number of trees per unit of land area. However, much elaborate pruning had to be done to achieve the relatively flat shape of the palmette and to incline the two trusses of the main scaffold branches (Fig. 1.6). The high cost of pruning was not compensated for by higher yields or better fruit quality.

Despite these changes in cultural practices and the few innovations introduced in pruning, the olive industry attracted few investments because the market for olive products remained relatively static until the late 1980s. Nowadays, economic

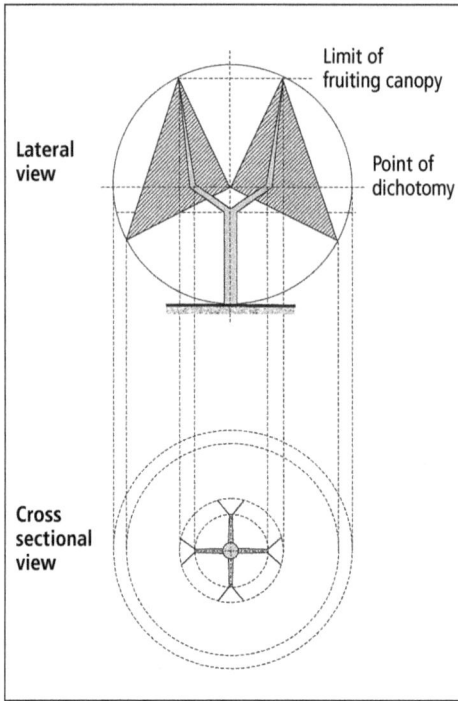

Figure 1.5 Schematic diagram showing the canopy profile of the polyconic vase generated by the rotation of a rectangular triangle around its right angle (modified from Roventini, 1936).

Figure 1.6 Young olive tree trained to palmette. The canopy is relatively flat and extends in the direction of the tree row. The two orders of main scaffold branches are being formed by inclination of lateral shoots at two different heights.

conditions are more favourable than in the past. In recent years, following the expansion of olive oil consumption and the change in dietary habits worldwide, there has been a new interest in olive growing. As a result, production has shifted gradually from traditional growing areas to others outside the Mediterranean basin. The development of a viable industry is taking place in countries where olive growing had never been economically relevant. Although this trend is still in progress, one can already foresee the implications for changes in cultural practices, including pruning strategies, and a demand for innovations that poses new challenges to scientists and growers. Large homogeneous plantations, often in flat lands, availability of water for irrigation, fertigation, highly mechanized orchards, and the absence of certain pests make these new areas strong potential competitors for the traditional olive-growing countries. This means not only that new methods of cultivation are being introduced, but also that some old concepts can be revived and adapted to meet the current needs of modern olive growing.

[1]The works of Davanzati, Soderini and Vettori were written in the sixteenth century but they were published later. The dates of the original manuscripts are reported in the text, although they are listed in the references with the year of the book editions that were actually read.

Chapter 2

Basic principles for pruning woody crops

2.1 General

"Many people cut, very few prune." This old saying well describes the fact that pruning is often misunderstood as a synonym for cutting. In fact, pruning covers a wider range of operations than just thinning or heading of shoots. Pruning is defined as the set of manipulations that can be used to form the structure of the canopy, influence the development of vegetative and reproductive organs, modulate growth of above- and below-ground parts, with the aim to maximize yield and fruit quality. Hence, pruning encompasses all those techniques whereby one can alter growth and development of the plant structures, including bud removal, heading, incision, inclination, girdling, twisting and so on. Practices like fruit thinning, irrigation, and fertilization are used in conjunction with pruning techniques to influence the rate of shoot growth or the relationship between growing organs.

Pruning is one of the most critical practices for a successful orchard. Pruning trees is not a trivial task, as it involves knowledge of the biology of tree species, of the interaction between the plant and the environment, as well as technical experience and economic wisdom. Pruning techniques vary with species. Although most principles and methods of pruning can be applied to all higher plants, in horticultural terms pruning is used almost exclusively for woody perennials. Trees are generally exposed to changing environmental conditions and extremes during their long life cycle. Because of their large size, processes occur in trees at a larger scale than in annual plants. An obvious example is the transport of sap from the root to the top of the canopy. The morphological and physiological features of different woody species can be quite distinct even in closely-related species. Pruning apricot trees is very different from pruning cherry trees; and pruning sour cherry differs from pruning sweet cherry, even though both species belong to the genus *Prunus*.

Within the same species the response to pruning may vary with cultivar, phenology, age, vigour, crop load and history of the plant. Two individual trees or branches are never perfectly identical, so pruning requires experience and open-mindedness. The array of possibilities and challenges offered by pruning trees is huge. Thus, one must learn the basic principles of pruning, as well as the specific characteristics of the crop, to be able to develop adequate pruning strategies.

The type of pruning also depends on the objectives of the grower. For instance, in pruning trees for wood production the main trunk is favoured rather than the lateral branches, whereas in pruning fruit trees the goal is to optimize fruit production. However, pruning for maximum yield or quality can often be so expensive that it is uneconomical. In this case, pruning techniques have to be adjusted to make them economically sound. Thus, the final decision about pruning should be based on economics as long as the physiology of the tree is respected. This compromise between economics and physiology is not always easy to achieve.

In order to prune trees correctly one must understand the relationships existing between growth and development of the different organs and the plant's response to environmental factors. Many examples exist indicating that similar plant responses can be obtained by different means and practices, or that the same manipulation performed at different phenological stages produces a different response. The main objective of pruning is to maintain an equilibrium between vegetative activity and reproduction so that yield potential is fully expressed.

Pruning is one of the most powerful tools for manipulating plant growth and reproductive processes. Pruning alters physiological relationships, the regulation of which can be quite complicated and multifaceted. Zucconi (1994) points out that pruning involves a higher level of complexity than do many branches of plant physiology and hypothesises that pruning is to plant physiology as pre-Galilean mechanics was to physics, when technical knowledge was more advanced than science. Indeed, most knowledge on pruning is merely empirical and comes from experience rather than scientific experiments.

The objective of this chapter is to provide the basic concepts of tree physiology and introduce the reader to the terminology of the pruning of fruit trees. The description of the biochemistry and physics of processes such as photosynthesis, carbohydrate partitioning, transpiration, flowering, nutrient uptake, nitrogen metabolism, and growth regulators, is beyond the scope of this work. These topics are well reported in many textbooks of plant physiology (Faust, 1989; Kramer and Kozlowski, 1979; Lambers et al., 1998; Salisbury and Ross, 1978).

2.2 The life cycle of the perennial plant

Trees are perennial plants that pass through several stages of development. During the juvenile phase the plant is unable to flower and bear fruit. Juvenile plants grow rapidly, produce long shoots with a high growth rate for a longer period of time than mature plants, and respond vigorously to cutting and manipulations. Juvenile leaf shape, size and colour can differ from those of mature foliage. Other minor traits, like leaf retention in deciduous species or the tendency to thorniness, are also signs of juvenility.

Once the juvenile phase is over, the plant attains reproductive maturity and bears fruits. Vegetative activity is reduced, but not suppressed. The canopy of some species becomes more open and expanded laterally, thus assuming the shape of a globe or an umbrella, whereas juvenile plants tend to grow more slender. The accumulation of dry matter and the response to cutting are also more attenuated than in juvenile plants.

In the third and final stage of the life cycle, the reproductive activity of the tree prevails over vegetative growth. Shoots grow at a slower rate, the increments in girth are decreasing due to the reduced activity of the cambium, the root system tends to expand in depth and shrink laterally, and yield declines. Short shoots are more numerous than long shoots. Trees can be brought back to conditions more typical of the mature stage by practices that favour vegetative growth (e.g. nitrogen fertilization, irrigation, heavy pruning).

2.3 The annual cycle of the perennial plant

The response to manipulations depends strictly on the phenological stage of the woody plant. Phenology is the study of the progression of physiologically distinct stages over the annual cycle, which unfold according to a genetically predetermined sequence under the influence of climatic factors. In perennial plants we recognise several phenological stages (see also Lang et al., 1987; Martin, 1987; Soule, 1985).

Budbreak indicates the opening of buds, after which shoots start elongating. The opening of flower buds may last from a few hours to several days depending on prevailing temperatures and species. When most flowers are open the plant is at full bloom. The percentage of flowers considered adequate for full bloom depends on the species (see Brief Glossary on page 18 for the estimation of full bloom in olive). Full bloom is followed by petal fall, after which the ovary enlarges, assuming that pollination has occurred. Fruit set occurs when the ovary reaches a diameter of about 4 mm. Final fruit set can be much lower than initial fruit set because of fruit abscission. In olive, fruit set should be assessed at least twice (60 days after petal fall and before harvest) to take into account the large number of fruits that abscise during the growing season. Apple trees show more definite periods of fruit abscission, whereas natural fruit drop is virtually absent in some species. Percentage of fruit set can be expressed in several ways, also depending on the flowering habit of the species. Fruit set (number of fruits divided by number of flowers) ranges from 0.15 to 0.3–0.8 in apricot and pear, respectively, whereas in olive this value is less than 0.05. Fruit set can reach values higher than one in certain species (e.g. cherry) if calculated as a percentage of inflorescences rather than of total flowers.

Fruit growth commences after fruit set and ends with harvest. Fruits are harvested at commercial maturity, which generally does not coincide with physiological maturity. In the initial stages of fruit development the fruit grows mainly by cell division, whereas the increase in size and biomass is due more to cell enlargement in subsequent stages. Massive sclerification of the endocarp (pit hardening) takes place between the cell division and cell expansion phases. During development the fruit undergoes morphological and biochemical changes, the most evident of which is usually the change in colour (veraison) due to chlorophyll degradation and accumulation of anthocyanins.

At the end of the growing season, deciduous trees drop their leaves and spend winter in a dormant stage. Although there is no apparent growth, metabolism does not stop during this period. In evergreen trees growth resumes promptly when temperature rises, and then flushes of growth occur during winter in mild climates. Nevertheless, even in evergreen trees like olive the chilling is needed to release flower buds from dormancy and to allow differentiation to proceed the following spring (see also section 3.6).

2.4 The balance between vegetative and reproductive activities

In fruit crops the establishment and maintenance of the balance between reproductive and vegetative activities are crucial for successful production. Errors in pruning or cultivation may delay the establishment of this equilibrium. Similarly, if external causes upset this balance the mature plant may return to an unproductive state. Unlike annual crops, in which crop load is directly proportional to the development of leaf area, in mature trees there is often an inverse relationship between leaf area and fruiting. This is due to the diversion of resources from reproductive to vegetative sinks when shoot growth and leaf expansion proceed at high rates.

The concept of balance between functions can be extended to plant parts and single organs. The root-to-shoot (R/S) ratio, which relates the dry matter of the above-ground parts to that of the root system, and the ratio between the mass of new wood (bearing leaves and fruits) to that of old wood (transport and mechanical support), are used as indexes of the relationship between functional parts and structural parts. An analogous index between structural and functional parts can be calculated for the root system, where the actively absorbing surfaces are present in the distal region and large roots are used mainly for transport and storage.

Trees with a similar R/S ratio can differ physiologically, and the measurement of masses *per se* is not indicative of the balance between functions within the tree. Typical indicators of altered R/S ratio are the rate of shoot growth, internode length, area of individual leaves, abundance of watersprouts and suckers.

An elegant and simple model describing the balance of functions in the woody plant was formulated by Zucconi (1994), who proposed that plant functioning is characterized by four different components: (1) the aerial factor is the set of compounds/functions that the root system is unable to produce/perform or would produce less efficiently than the canopy; (2) the radical factor is the exclusive or preferential product of the root system; (3) both factors are necessary to all plant parts for energy, structural materials and regulation; (4) both factors must be transferred from the site of production to that of utilization to be effective. The two basic principles of this model are that: (a) each factor is limiting for the organ or plant part that does not produce it, so that the activity of an organ is proportional to the limiting factor; (b) every contraction or expansion of the activity of an organ induces parallel changes in the activity of other organs and the whole system.

In perennial plants the balance between vegetative and reproductive activities is further complicated by apical dominance (see section 2.7) and carry-over effects due to the plant's history. Historical effects are inherent to the plant and can be carried over long periods of time. An example is the phenomenon of alternate bearing, which becomes more prominent with age.

2.5 Autonomy of plant parts

A peculiar feature of plants is that there is a certain degree of independence between various organs and parts. Unlike higher animals, in which all functions and processes are under strict control of the nervous system, functions are less integrated in plants. The degree of autonomy of parts is more evident in perennial woody species than in annual plants, and particularly in large trees. Consequences of autonomy are that some branches often overgrow others, or that one branch or sector of the canopy is out of phase for fruit production with the rest of the tree. Hence, local responses are important in understanding the tree response to manipulations such as pruning.

The root system also shows a partially autonomous behaviour. Root density is high in fertile humid soils and low in arid or poor soils. A common example of an uneven root system is given by the increase in root density near the dripper in irrigated orchards at the expense of lateral root expansion. This situation can result in poor anchorage and more sensitivity to drought or lack of nutrients if the irrigation system fails for even very short periods during the dry season.

2.6 Photosynthesis and light interception

Photosynthesis is the process whereby carbon dioxide present in the atmosphere is assimilated by the leaf in the presence of light, chlorophyll and enzymes. Carbon dioxide penetrates by diffusion through the stomata (pores present on the leaf surface) and is incorporated into carbon products via a carboxylation reaction. Carbohydrates and oxygen are the main end products of the photosynthetic reaction; the former are accumulated in the leaf cells or transported to other tissues and organs where they are utilized or stored, whereas oxygen diffuses back into the atmosphere. The biochemistry and physics of the photosynthetic process are well described in many textbooks of plant physiology and are not discussed here.

Carbohydrates are compounds where most energy is stored. When broken down by respiration, carbohydrates supply carbon units and energy needed for plant metabolism. Leaves are the main photosynthetic organs of a tree; the contribution to photosynthesis by other green parts is minor and can be assumed to be negligible. For the whole plant, the amount of carbon assimilated through photosynthesis can be estimated by the following equation:

$$A = \alpha \times \beta \times Q - R$$

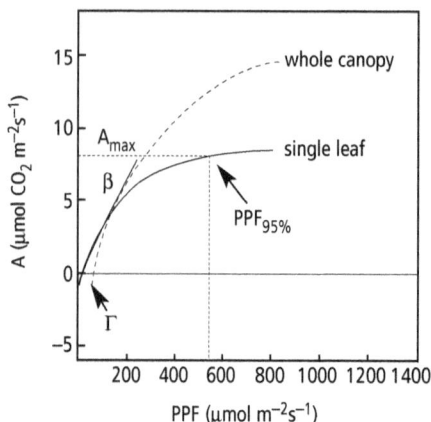

Figure 2.1 The relationship between net photosynthetic rate (A) and incident photosynthetically active radiation (PPF) for a single leaf and the whole canopy. The dotted line indicates the radiation level at which photosynthetic saturation occurs for single leaves. Note that photosynthesis of the whole canopy does not reach a saturation threshold, as it continues to increase with increasing PPF. The CO_2 compensation point (Γ), which is the light level at which photosynthesis equals respiration, and the apparent quantum yield (β), estimated as the slope of the linear part of the curve for single leaves are also indicated.

where A is net assimilated carbon, Q is radiation available from sunlight, R is the respiration rate, α and β are coefficients expressing the percentage of available light intercepted by the canopy and the yield efficiency of converting light into chemical energy, respectively. Since β, the quantum yield efficiency, and R are fairly constant across woody crops, main differences in assimilated carbon lie in α, the amount of light intercepted by the canopy. In turn, light interception is positively correlated with leaf area development, which, in discontinuous canopy systems like trees in an orchard, is often the limiting factor for biomass production per unit land area. Since canopy volume and shape strongly influence the amount of intercepted light, the training system may affect plant productivity and efficiency (Jackson, 1980).

The process of photosynthesis is driven by light. The relationship between photosynthetic rate and incident photosynthetically active radiation (in the range between 400 and 700 nm) is shown in Fig. 2.1. Carbon assimilation of individual leaves increases with incident radiation up to a saturation level, beyond which it stabilises because biochemical limitations occur. The radiation level for photosynthetic saturation of the whole canopy is much higher than for individual leaves, and especially so in dense canopies where most leaves are shaded and photosynthesise at much lower rates than their maximum potential (Fig. 2.1).

In fruit trees, the relationship between the amount of intercepted light and fruit production is not a simple one. In fact, carbon not only has to be assimilated through photosynthesis but also has to be transported to the reproductive sinks. The ratio between the dry matter accumulated into the usable part of the plant product (fruit for fruit trees) and total dry matter is defined as the harvest index. The importance of the harvest index, which represents a simple way to measure dry matter allocation toward the economic objective of cultivation, can be fully understood when considering that most of the increase in yield of wheat cultivars achieved in the last century has been obtained by breeding for higher dry matter allocation to the seed rather than for carbon assimilation in itself (Gifford et al., 1984). In fruit trees, the harvest index is quite small in comparison with that of more productive annual crops or perennial crops used for wood or fibre.

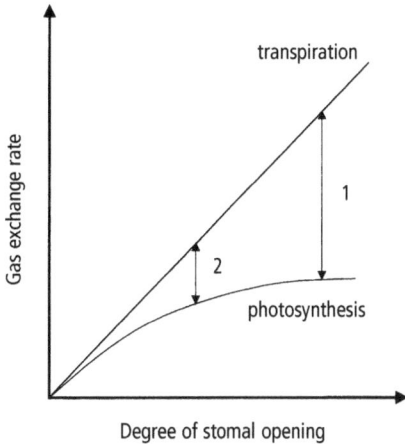

Figure 2.2 The compromise between transpiration and photosynthesis expressed as a function of stomatal opening in a hypothetical leaf. Biochemical limitations prevent further increase of net photosynthetic rate when stomata are fully open, whereas transpiration rate continues to increase (1). When stomata are only partially open more carbon is accumulated per unit of water vapour transpired (2). Note that transpiration is about three orders of magnitude greater than carbon assimilation rate (not drawn to scale).

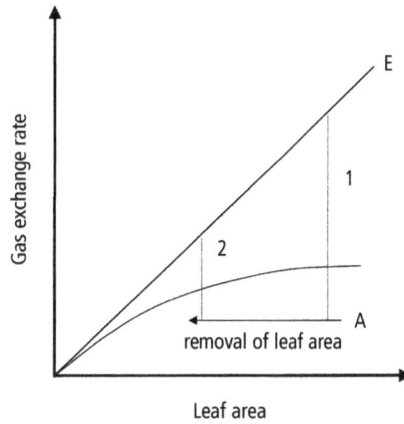

Figure 2.3 Diagram showing the effect of pruning on the balance between photosynthesis (A) and transpiration (E) of the entire tree. Removal of leaf area from level (1) to level (2) increases the water use efficiency of the tree. Not drawn to scale.

Most pruning operations involve removal of photosynthetic area. The decrease in leaf area reduces the capacity to assimilate carbon and the vigour of the remaining shoots. If this reduction is excessive in relation to the remaining leaf area, the amount of available energy in the branch or the tree may become limited. An excessive elimination of foliage by pruning can slow down growth during the training phase when the initial leaf area is small (see also Chapter 5). Therefore, removal of leaf area from an individual shoot or branch can be used to weaken a single shoot or to balance the vigour of scaffold branches.

Photosynthesis occurs at the expense of water loss. Carbon is assimilated concomitantly with transpiration through stomata, so that for each unit of carbon acquired, about a hundred times as much water is lost. As stomata open, both photosynthesis and transpiration increase. However, beyond a certain threshold of stomatal opening, photosynthesis does not continue to increase because biochemical limitations overcome the physical limitations to gas diffusion via stomata whereas transpiration continues to increase (Fig. 2.2). As a result, the highest efficiency for plant assimilation is reached when stomata are not fully open because the unit of carbon gained costs less water. Hence, removal of leaf area by pruning reduces water consumption and increases the water use efficiency (ratio between net photosynthetic rate and transpiration rate) of the canopy (Fig. 2.3).

2.7 Apical dominance and response to inclination

In higher plants apical dominance indicates the control exerted by an apex over the formation of lateral buds and shoots. Dominance can be either complete, totally suppressing the growth of laterals, or partial. In the latter case the apex influences not only growth but also the angle of insertion and orientation of branches (Martin, 1987). Apical control can be strong over the lateral buds and shoots of an individual axis, but weak over the whole tree. Most fruit tree species are polyaxial and competition among axes prevents a single stem from prevailing over others. In most cases, a central leader can only be obtained by eliminating the competing lateral shoots. However, once pruning is interrupted, the central leader is lost and the tree returns to a polyaxial structure.

Two major patterns of lateral shoot growth can be distinguished in fruit trees. Acrotony refers to the control by the apex over lateral buds and shoots according to a gradient from top to bottom. On the contrary, a basitonic behaviour indicates a prevalence of branching at or near the base of the shoot, with shoots in the proximal part of a branch overgrowing those inserted in the distal part.

Inclination is the altering of the angle of the axis of a branch, or shoot. It is a simple operation that can affect the growth rate of a lateral structure and increase its reproductive activity. The wider the angle from the vertical axis, the stronger the reduction of vegetative growth of the inclined shoot. In basitonic species (e.g. olive) inclination of a shoot further increases the prevalence of shoots in the basal part of the main axis (Figs 3.8; 3.9). Inclination beyond the horizontal (bending) causes an even stronger reduction of vegetative growth of the main axis. Bent shoots are usually shortened the following year by heading back. Inclining watersprouts when they are still tender favours their development into flowering shoots and reduces crowding in the interior of the canopy.

Another consequence of inclining a branch is the production of additional wood (reaction wood), which tends to counteract the force applied to incline the structure. Reaction wood is present naturally as a swelling on the lower side of the insertion point of a lateral shoot (Kramer and Kozlowski, 1979). The border area between the main axis and the lateral branch is identified by the presence of a branch collar on the lower side of the point of attachment, and a corresponding area on the upper side (branch bark ridge). These areas are used to determine the point at which lateral branches are cut (see also section 4.3).

2.8 Cutting

Cutting is the most commonly used operation in pruning and is often confused with pruning itself. Cutting is used to reduce competition between shoots, stimulate vegetative growth, reduce photosynthetic area, and improve light penetration. Cutting usually stimulates growth of the parts remaining after the cut, but the effect differs depending upon whether it is used to eliminate shoots or shorten them. There are two distinct types of cuts that can be used.

Thinning is the elimination of shoots or branches from their point of origin, or of the main axis near a lateral shoot that subsequently assumes the terminal role. This lateral shoot reduces the tendency of lateral buds to develop. Removal of entire shoots or branches stimulates the vigour of those that are left by reducing the competition for light, water, nutrients and carbohydrates, although the total amount of growth is reduced.

Heading consists in cutting young (current season or one-year-old) shoots back to a bud, or shortening an older shoot or branch back to a lateral too small to play the terminal role. Heading cuts made at the same height from the ground or along the side of the canopy are commonly termed "topping" and "hedging", respectively.

The effect of thinning is more uniformly distributed over the remaining shoots, whereas heading causes a stronger response next to the cutting point. The more severe the pruning, the more the remaining individual shoots will be invigorated, but the less the total growth of the pruned branch or tree will be. Shoots can be cut at different points: apical, median, or basal. The lower the cut, the more vigorous the growth of the shoots developing from near the cut surface. In all cases the cut is made above a bud from which a new shoot will emerge.

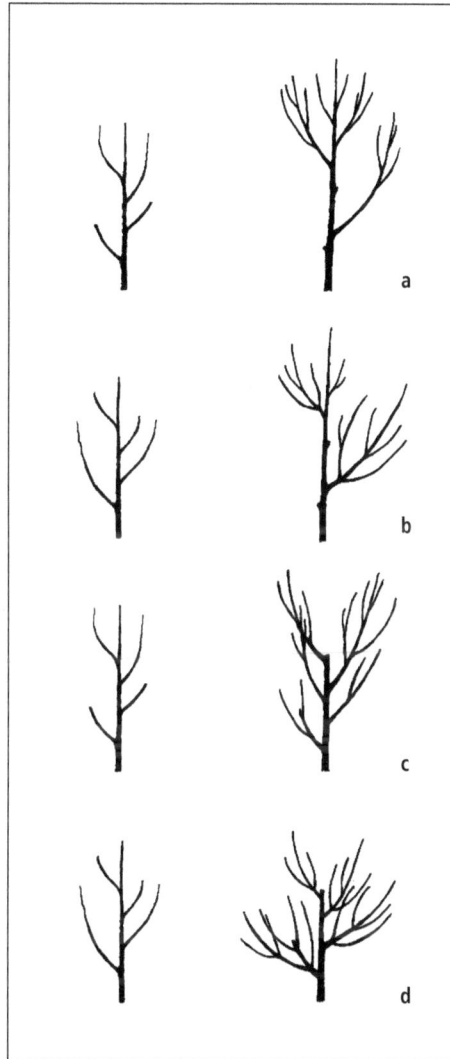

Figure 2.4 The effect of cutting at the beginning of the growing season on growth of acrotonic (a) and (c), and basitonic (b) and (d) species. (a) and (b) removal of entire shoots; (c) and (d) heading of the main axis and lowermost lateral shoots.

The pattern of vegetative response to shoot heading or thinning differs in acrotonic *versus* basitonic species. In the former species (e.g. peach, apple), heading stimulates growth of the lower (more basal) buds and shoots, whereas in the latter (e.g. olive) growth of shoots close to the cut is favoured. Therefore, heading results in an attenuation of the growth pattern of plants (reversing to some extent the natural gradient in growth between apex and base), but at the same time stimulates growth (Fig. 2.4).

2.9 Other pruning operations

Incision, notching, girdling, inverted girdling, and twisting are examples of the many options other than cutting and inclination that can be used to stimulate budbreak at a particular point, to promote shoot growth, or more generally, to alter the relative vigour of entire trees or parts thereof.

Girdling is the removal of a ring (5–10 mm wide) of bark, without affecting the xylem, to improve flower bud development, bloom, fruit set and fruit size and to reduce vegetative growth above the point of girdling. Girdling is usually done with a grafting knife and the girdle can be simply covered with a PVC strip. Girdling results in beneficial effects only on plants with sufficient vigour; under poor conditions for growth, debilitating effects can override any benefit.

The effect differs according to the period when the shoot or branch is girdled. Lavee et al. (1983) showed that girdling in winter months increased yield of individual olive branches of cultivars 'Uovo di Piccione' and 'Sde Eliyahu', whereas late summer or autumn girdling had no effect. Spring girdling also increased yield, although to a lesser extent, but girdling in the early summer caused a decrease in yield in both cultivars (Lavee et al., 1983). Lòpez-Rivares and Suarez (1990) showed that girdling increased fruit size only if done 30 days before full bloom under irrigated conditions. The effect of girdling has also been studied in relation to alternate bearing. Girdling increased flowering and fruit set of individual branches of cvs. 'Uovo di Piccione' and 'Manzanillo'; the non-girdled branches also had a lower percentage of perfect flowers (Ben-Tal and Lavee, 1984). The same authors reported a reduction in biennial bearing over four years of study in both cultivars; this was more evident in 'Uovo di Piccione' than in 'Manzanillo'.

In scoring, an incision in the bark is made with the point of a knife, a saw or a chain saw in spring. Unlike girdling, the cut is made for only one-quarter to one-half of the circumference. The width of the wound (2–4 mm) must be adjusted to the size of the branch and must allow the bark to heal rapidly. The incision stimulates bud growth below the point of cut, but the effect is confined to the year when it is performed. In olive the incision can be used to stimulate budbreak and shoot emergence at specific points on woody branches. Notching and twisting are seldom used. Twisting is used, as an alternative to cutting, for weakening watersprouts in vigorously growing plants.

Note that all these operations are rarely used in commercial groves as they are too elaborate and time-consuming to be recommended at the orchard level; however, they can be useful for research purposes or to prune olive trees in gardens and hobby orchards. Their correct performance requires experience and the reader should apply these techniques only after sufficient practice.

Brief glossary

General

Alternate bearing

Sequence of a high yield in one year followed by a low yield in the subsequent year. In woody perennials alternate bearing can be evident at different levels (branch, tree, orchard); its occurrence increases with plant age. Although the causes and regulation are endogenous to the plant, climatic events (frost, drought) often trigger the onset of alternate bearing. Typical alternate bearing species are apple, citrus, olive, pecan and pistachio.

Anthesis

Opening of flowers with anthers and ovary exposed. Flower buds develop into flowers under a temperature stimulus.

Apical dominance

Control by an apex on growth and development of lateral meristems. Apical dominance proper is the influence of the apex over lateral buds on a single branch, whereas apical control refers to the influence of the main growing point on all branches of a perennial plant (Martin, 1987).

Basitony

The tendency of shoots to develop at or toward the base of the stem. In basitonic species basal shoots often overgrow the central axis and apical lateral shoots.

Dormancy

Temporary interruption of visible growth of any plant structure containing a meristem. Physiologists distinguish between different types of dormancy depending on the receptor organ of the environmental or endogenous signal. Lang et al. (1987) define: *endodormancy* if the specific perception occurs within the affected structure of the plant; *paradormancy* if it originates in a structure other than the affected structure; *ecodormancy* to include all causes of dormancy due to unsuitable environmental factors which are non-specific in their effect on plant metabolism (e.g. temperature extremes, nutrient deficiency, water deficit). Bud dormancy is considered released when 25 to 75% (more typically 50%) of the buds are capable of growth.

Drupe

A fruit formed by a thin membranous outer layer (epicarp), a fleshy mesocarp and a stony endocarp. Apricot, cherry, peach, plum and olive are also called stonefruits because of the presence of the stony endocarp.

Flower bud induction

First stage of the reproductive cycle. The first stimulus for flower bud formation is perceived (not necessarily by the bud itself), but no visible or biochemical signs can be detected in any plant tissue. In olive, flower bud induction occurs in the summer.

Flower bud initiation

The first stage when flower buds can be identified by histochemical or biochemical tests. In olive it occurs between the end of the summer and autumn.

Flower bud differentiation

Development of flower parts following initiation. In most fruit trees of the temperate zone, flower buds are fully differentiated before winter, but in olive differentiation occurs during or at the end of winter. A period of chilling temperatures is needed to proceed irreversibly from flower bud induction to differentiation.

Full bloom

Indicates the stage when the majority of flowers are open. For specific shoots full bloom is the first day on which more than 50% of the flowers are open on at least 70% of the inflorescences of the shoot (Rapoport and Rallo, 1991). For whole trees full bloom is the midpoint between the start of full bloom (the first day when open flowers are at their most abundant as a stage of the inflorescence) and the end of full bloom (Barranco et al., 1994).

Fruit set

Initial enlargement of the ovary following fertilisation expressed as a percentage of the total number of flowers. In olive, final fruit set is less than 5% of the total number of flowers.

Maturation

Final stage of fruit development. In the olive drupe it is characterized initially by the slight discoloration of the epicarp (green maturation) followed by veraison, and later by the accumulation of oil, polyphenols, and flavour compounds.

Phenology

The study of the progression of physiologically distinct stages over the annual cycle. Phenological stages unfold according to a genetically predetermined sequence, which is under the influence of climatic factors.

Pit hardening

Stage of fruit development when endocarp sclerification occurs to a large extent. In olive, oil accumulation begins after the end of the pit hardening stage.

Veraison

Initial change of colour of the fruit from green to final colour due to chlorophyll degradation and accumulation of anthocyanins.

Pruning terminology

Arrow

Terminal part of the main axis of a shoot, branch, or tree.

Bending

Inclination below the horizontal axis.

Burr-knot (ovulus)

Hyperplastic tissue with high meristematic activity usually formed on the crown of the olive plant below the soil line. Burr-knots were once used routinely to regenerate whole plants.

Crown

The part of the plant connecting the root system with the trunk.

Cutting

Includes both thinning and heading (Harris, 1994). Thinning is the elimination of shoots or branches from their point of origin, or of the main axis near a lateral that assumes the terminal role. Heading consists in reducing the length of young shoots back to a bud, or the shortening of a branch back to a lateral structure too small to play the terminal role. Pinching, tipping, heading, and stubbing indicate shortening of shoots in increasing order of severity, that is, with increasing amounts of wood being removed. The vegetative response consequent to thinning is distributed quite evenly over the remaining shoots, whereas heading causes a strong response next to the wound.

Feathers

Lateral shoots developing from axillary buds formed in the same year. Feathers are also called prompt shoots.

Girdling

Removal of an annular strip of bark without affecting the xylem. It is used to reduce vegetative growth above the point of girdling and to stimulate flower bud development, fruit set and fruit growth.

Hedging

Uniform cutting of the lateral part of a canopy. It is usually done mechanically by rotating blades or sickle bars carried on a tractor.

Inclination

Altering the angle of the axis of a shoot or branch. Inclination is used to slow down growth of the main axis, and to stimulate flower bud development and the rise of vegetative buds and shoots in the proximal part of the inclined structure. In olive, inclination should be used carefully because it induces vigorous growth of basal shoots on the upper side of the branch.

Notching

Cutting of a small portion of the bark and cambium just above or below a bud. In the former case notching stimulates vegetative growth of the bud; in the latter it promotes flower bud initiation.

Pinching

The removal of the distal part of a current season's shoot before lignification to stimulate emergence of new lateral shoots. Since the tissues are tender, pinching is done manually.

Pruning

Set of manipulations that are used to form the structure of the canopy, influence the development of vegetative and reproductive organs, and modulate growth of above- and below-ground parts to maximize yield and fruit quality.

Scoring

Superficial cutting of the bark only for part of the circumference. Scoring is used to stimulate bud growth below the point of the cut and to produce new shoots at specific points of the woody branches. The effect is limited to the growing season when it is performed.

Shoot

    Vegetative shoots have long internodes, do not bear flower buds, and usually grow erect on two-year-old wood. *Fruiting* shoots bear only flower buds; they are thin with short internodes and located at the end of fruiting branches. *Mixed* shoots are fruiting shoots with flower buds in the axils of leaves and the terminal bud(s) vegetative.

Sucker

    Suckers are vigorous vegetative shoots originating from the crown of the plant.

Topping

    Uniform cutting of the apical part of a canopy at the same height from the ground. Topping is usually done mechanically as in hedging.

Twisting

    Torsion of vigorous shoots to reduce their vegetative growth.

Watersprout

    Vigorous vegetative shoot with long internodes. Watersprouts grow almost vertically, on the upper side of horizontal branches in inner, shaded parts of the canopy and in the tops of trees that have been headed back.

Chapter 3

Physiological background for pruning olive trees

3.1 General

Pruning olive trees is neither easier nor more difficult than pruning other fruit trees. It is just different. There are many distinct physiological features that make the olive plant unique. The basitonic growth habit, the response of shoots and branches to inclination, the high capacity of resprouting, and the reproductive biology are some examples of physiological features peculiar to the olive plant. Therefore, pruning techniques used for other species are not necessarily appropriate for olive.

This chapter focuses on the basic biological processes of the olive plant that must be considered to understand pruning. Pruning techniques have been developed almost entirely empirically; only in the last century have physiological concepts been applied. To optimize pruning efficiency today, pruners, especially beginners, must become familiar with the specific features of the olive plant and with the major responses to manipulations.

Figure 3.1 Old olive tree in a traditional grove.

3.2 The life cycle

Olive is a long-lived species. There are many records of trees several centuries old (Fig. 3.1) and some whose age has been estimated to be over 2000 years. The most famous examples are the eight olive trees in the Garden of Gethsemane in Jerusalem, presumably dating back to Christ's lifetime. Life is shorter in cold climates; whole trees seldom survive for more than a couple of centuries, but, because of the high capacity to regenerate the canopy from suckers, plants often have an ancient root system and a young canopy formed by multiple stems (Figs 3.2; 3.7; 8.17).

In modern olive growing the time frame over which economic analysis should be projected is about 40 years. However, old trees can be very productive despite their age. Under many circumstances

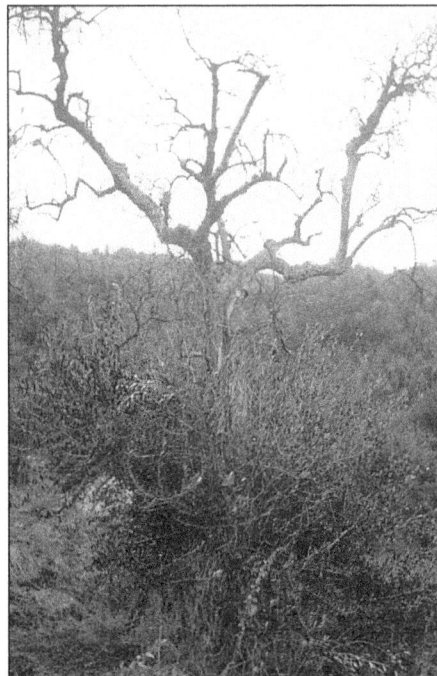

Figure 3.2 Suckering of an old olive tree killed by frost.

they continue to produce high yields once they are successfully rejuvenated by pruning (see section 6.8). The main constraint to rejuvenating old trees is not their future recovery or productivity, but rather the high cost for pruning such large canopies (Fig. 3.3). Hence, rejuvenation of old trees is often uneconomical.

In olive the juvenile phase can last several years. Seedlings usually require up to 10 years before they start flowering. However, under highly inductive conditions (growth concentrated on a few apices, high fertilisation and irrigation regimes, sunlight supplemented with artificial light during winter) the first flowering can occur at the end of the third year after seed germination (Lavee, 1990). Refinements of this protocol have been reported by Rallo (1999), who induced 5 to 40% of plants (depending on the cultivar) to flower 30 months after germination.

The trees propagated from mature tissue are much more precocious than seedlings. Moreover, by the time plants are ready for sale they are already 12- to 18-months-old. If olive trees are correctly managed in the field they will flower and begin production in the third or fourth year after planting; however, climatic conditions or erroneous practices (e.g. excessive pruning), that stimulate vegetative growth rather than reproductive activity, can delay flowering.

The juvenile plant usually produces several vigorous upright shoots. The juvenile shoot can be identified by morphological characteristics, including growth rate and internode length. Because of the high growth rate of these shoots, the points of insertion of the opposite leaves of each node are further apart. Juvenile shoots often

Figure 3.3 The canopy of this olive tree (cv. Koroneiki) has been rejuvenated and reduced in size by pruning. The tree has abundant foliage for bearing heavy crops.

bear three leaves per node rather than the typical two of mature shoots (Fig. 3.4a). The leaves are smaller, thicker, greener and rounder than mature leaves (Fig. 3.4b). The foliage of mature plants assumes the characteristics typical of the cultivar. The canopy of the mature plant is broader than that of the juvenile plant (see also section 3.3).

Old trees have a high ratio of wood to foliage than do young trees (Figs 1.2; 3.1; 4.8). Their potential reproductive activity is high, but, because of little vegetative growth, old trees generally yield less than mature ones. The pruning of old trees is designed

Figure 3.4 (a) Morphological characteristics of mature (left) and juvenile (right) shoots. Note the different sizes, shapes and numbers of leaves on each node; (b) leaves from mature (above) and juvenile (below) shoots.

mainly to stimulate vegetative growth and reduce tree size by making few large cuts on major branches (see also sections 6.7 and 6.8).

3.3 The growth habit

The growth habit of the olive plant is quite distinct from that of other fruit trees. If left unpruned, the plant tends to grow as a bush rather than as a tree, with the canopy often formed by multiple stems originating directly from the crown. Even in a seedling the central stem does not grow vigorously and is progressively overgrown by lateral shoots. The taproot is replaced by lateral roots often originating from burr-knots (ovuli), hypertrophic tissue of high metabolic and meristematic activity present a few centimetres below the soil line (Fig. 3.5). Burr-knots can produce a whole tree if removed from the mother plant.

Figure 3.5 Burr-knots present on the crown of a young olive plant. The crown has been excavated to expose the knots present below the soil line (courtesy of Dipartimento di Coltivazione e Difesa delle Specie Legnose, University of Pisa).

The number of suckers originating from a stump after the canopy is cut off is extraordinarily high. Over 300 new suckers can emerge from 30-year-old stumps of about 1.5 m circumference. The self-thinning of newly-formed suckers is also exceptional; their number decreases to about 100 and 50–40 by the end of the first and second growing seasons, respectively (Cantini et al., 1998; section 8.7).

Figure 3.6 The natural growth habit of olive plants (cv. Mignolo) growing in the field. (a) three years after planting; (b) six years after planting. Trees were unpruned.

Figure 3.7 Growth of plants resprouting from the crown of trees whose trunks have been cut at ground level at the beginning of spring. The plants shown are at the end of the first (a), second (b), or third (c) growing season.

Figure 3.8 The natural growth habit of a vertical shoot at the end of the first growing season. The basitonic pattern of growth is evident.

The canopy profile of unpruned plants is approximately spherical, hemispherical or ellipsoidal. The canopy tends to expand laterally more than in height if left free to grow for a few years (Fig. 3.6). Similar contours are obtained when suckers emerging from the crown after the canopy is cut are allowed to grow without pruning (Figs 3.7; 8.17).

The shape of the olive plant is the result of both the basitonic growth habit of shoots and the tendency to produce many stems from the adventitious buds. The profile of a vertical shoot growing undisturbed is an asymmetrical diamond (Fig. 3.8). The lateral shoots in the proximal part grow more than do those in the distal part. Similarly, shoots on the lower side of a branch are less vigorous than those growing on the upper side (Fig. 3.9). In basitonic species the response to inclination further emphasises the tendency of basal shoots to prevail over apical ones. The dominance of the main axis is progressively weakened from the base to the terminal end by competition with lateral shoots. Basal shoots inserted on branches inclined at over 45° from the vertical axis tend to grow upright and may become so vigorous as to suppress the leader completely (Fig. 3.9).

Figure 3.9 The pattern of growth of lateral shoots inserted on an inclined axis. The basal shoots on the upper side grow more vigorously than the basal shoots on the lower side or those on the apical part of the main stem. The shoot was photographed with the same inclination it had originally in the canopy.

The basitonic habit and the response to inclination of olive have important practical implications. First, unlike peach, cherry and other tree fruit species in which the supremacy of the central leader is more easily maintained, the central axis of olive plants does not grow as vigorously as do lateral shoots. Thus, repeated thinning of the uppermost lateral shoots and elimination of vigorous laterals inserted in the lower part of the stem are needed to maintain the central leader in most olive cultivars. This holds true for both the single vertical axis of central leader systems and the primary branches of the polyconic vase (Figs 1.4; 5.5; 8.10). This is the reason why in central leader systems the lateral shoots growing next to the terminal end of the stem (arrow) must be eliminated periodically by pruning. Second, shoots developing on the basal upper side of inclined branches often grow so vigorously that they become watersprouts, and will have to be removed by pruning. The dramatic response to inclination explains why training systems requiring extensive use of this practice (e.g. palmette, Y-trellis) fail when applied to olive: the cost of inclination plus elimination of watersprouts is not compensated for by any beneficial effect on yield or fruit quality (see also Chapter 1). Inclination was once widely used to train the primary branches of vase systems (Breviglieri, 1961; Roventini, 1936). To reduce costs the angle of inclination of the main scaffolds is now obtained by cutting. This angle is progressively widened by cutting next to a lateral shoot with a wider angle of inclination than the removed part. The lateral structure thus becomes the terminal of the branch (Fig. 5.5; section 8.1).

Olive plants can be pruned to shapes different from their natural growth habit. By doing so, canopy contours can vary from bowl-shaped to triangular, rectangular and so on. A single trunk can also be developed by appropriate pruning (see Chapters 5, 6 and 8). Although the natural shape of the canopy can vary somewhat from individual to individual, the growth habit is a cultivar-specific characteristic. Four main habits can be identified: upright or erect; open or round; semi-pendulous; and pendulous. Most cultivars show an open habit of growth, which is intermediate between the pendulous and erect habits. Cultivars with an open canopy are: Arbequiña, Ascolana Tenera, Giarraffa, Hojiblanca, Leccino, Manzanilla, and Picual. The cultivars Barnea, Carolea, Mission, Moraiolo and Uovo di Piccione are upright, whereas Coratina, Correggiolo, Frantoio, and Koroneiki are semi-pendulous. The pendulous type of growth is not very common. 'Maurino' and 'Pendolino', used as pollinators for many Italian cultivars, are pendulous.

Yet, differences in appearance of the canopy can arise even between cultivars with a similar growth habit. For instance, the canopy of 'Coratina' trees is less orderly than that of 'Frantoio' trees, despite the similar semi-pendulous habit of both cultivars. The canopy of 'Maurino' trees is compact and dense, whereas that of 'Pendolino' is rather sparse. The difference results from differences in growth tendency (excurrent or outward in 'Pendolino' versus decurrent or inward in 'Maurino'). Excurrent or decurrent forms are the result of the duration and intensity of dominance exerted by the apex over the lateral shoots (Bongi and Palliotti, 1994). The 'Moraiolo' has a rather peculiar canopy; it is easily distinguishable from other cultivars with an erect habit, because of the clumped distribution of foliage around the terminal end of erect shoots. Identification of growth habits is important for classifying cultivars, but also for proper pruning (see also sections 4.3, 5.6 and 6.2).

3.4 The evergreen foliage

Olive is an evergreen species that retains leaves all year round. Leaves persist for two to three years. However, if trees/branches are shaded or under stress conditions, leaves last less than two years. From the point of view of energy, the olive leaf is more expensive than that of deciduous fruit trees. This is partially due to the evergreen character of the leaf and the abundance of compounds (waxes, celluloses, carbohydrates, lignins, phenolic substances, and oils) in the cuticle, cell wall and cytoplasm of cells. Most of these compounds serve mainly for protection against biotic and abiotic stress. The investment is partially compensated for by the longevity of the olive leaf and by the greater annual photosynthetic gain. The cost of maintenance appears to be less in sclerophyllous evergreens than in deciduous species (Meriño, 1987).

Because of their evergreen character olive leaves photosynthesise throughout the annual cycle. The rate of photosynthesis is optimal at 25–28°C with drastic reductions below 5°C and above 35°C. The annual photosynthetic balance includes reduction in assimilation due to temperature limitations in winter and summer (Gucci, 1998b). Further limitations to carbon assimilation result from soil water deficit during summer drought (Angelopoulos et al., 1996). Photosynthetic rates are not high in olive because of both leaf mesophyll and gas diffusion limitations. Mesophyll limitations can be ascribed to anatomical and biochemical factors, such as thick cell walls, low surface of mesophyll cells exposed to the intercellular solution, and low protein content including that of the Rubisco enzyme responsible for the carboxylation reaction in the chloroplast. In addition, the hairiness of the leaf surface and the presence of trichomes limit the leaf boundary layer.

The small size and rigidity of the olive leaf allow efficient exchange of heat when the temperature is high in the summer. The presence of trichomes seems to protect stomata and mesophyll from UV radiation which is high in Mediterranean climates (Grammatikopoulos et al., 1994). The threshold for light saturation has been estimated to be 500–800 μmol m^{-2} s^{-1} of photosynthetically active radiation (Bongi et al., 1987; Gucci, 1998a; Tombesi, 1988).

Factors altering the source-sink relationships of the plant are reflected in shoot growth patterns and the status of root carbohydrate reserves. The most active sink in the fruiting shoot or branch is the fruit, which reduces considerably terminal elongation and dry matter in the leaf or shoot (Rallo and Suarez, 1989). Fruiting or the removal of the canopy also reduce the starch content of small roots (5–15 mm in diameter) (Gucci R. and Cantini C., unpublished results).

3.5 Resprouting and response to cutting

Olive has a high capacity to regenerate a new canopy from the numerous adventitious buds and burr-knots at the base of the crown if the top is damaged, removed by pruning, or stressed. The growth often assumes a disordered pattern, with many growing points all competing with one another. The excessive crowding of shoots emerging after a damaging event makes the thinning of new shoots essential. However, the shoots should be selected only when the new canopy has grown sufficiently. Thinning of suckers will be done lightly only after the onset of a balance between the above-ground parts and the root system, which usually occurs in the third or fourth year after removing the canopy (see also sections 6.9 and 8.7). The growth in the three years following cutting the trunk can be extremely vigorous. In a trial where the orchard was managed by coppicing, total shoot elongation (including lateral ramifications) was estimated to be over 400 m per plant within the first growing season after resprouting (Fig. 3.7).

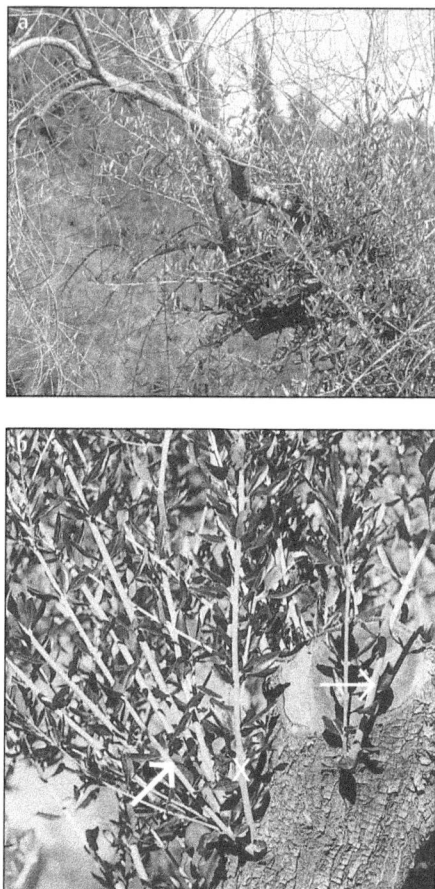

Figure 3.10 (a) Disorderly pattern of resprouting from woody tissue after cold damage; (b) new shoots originating from adventitious buds in response to elimination of a major branch. Note the crowding of shoots growing from near the cut surface. The shoot marked X will be eliminated by pruning, whereas the arrows indicate shoots suitable to reform the main axis of the primary branch (on the right) and a secondary branch (on the left).

The response of olive to cutting is typical of basitonic species (Fig. 2.4b, d). Shoots developing after cutting originate from axillary buds; each node bears buds in the axils of opposite leaves so that two shoots sprout from each node. In addition to axillary buds, olive has adventitious buds, which have a high sprouting capacity (Figs 3.2; 3.10). Adventitious buds can be found on virtually all vegetative parts of the plant; they are not evenly distributed and are usually invisible to the naked eye because they form under the bark. Adventitious buds may remain inactive for years until they receive a stimulus to break. Cutting, damage, or stresses (low temperature, waterlogging, pests and diseases) that weaken or kill the distal part of the branch can stimulate the growth of such buds.

There is no orderly pattern of budbreak from adventitious buds (Fig. 3.10a); consequently, the population of shoots developing from them must be adequately thinned. Figure 3.10 shows an example of how to proceed in the selection of shoots arising after cutting a large branch. The shoot to be removed (X) is growing almost vertically and is too vigorous in comparison with others developing from the same group of buds. It would also compete strongly with the shoot growing closer to the cut surface (indicated by the arrow on the right), which is in a better position to form the terminal of the primary branch. The shoot indicated by the arrow on the left is better suited to form a secondary branch and will be favoured by eliminating at least 50% of the neighbouring shoots. The thinning of shoots arising after a large cut should be carried out over two or three growing seasons to allow a more gradual development of the remaining shoots.

Classification of olive buds and shoots

Buds of the olive plant can be classified according to various criteria (Morettini, 1972). Classification based on position distinguishes terminal (apical), axillary (lateral), and adventitious buds. Terminal buds are formed at the apex of the shoot and they are usually vegetative. Axillary buds develop in the axils of leaves; each node bears two opposite buds. Axillary buds can produce either lateral shoots or inflorescences. Adventitious buds are formed on parts of the plant other than the leaf axil or the apex. Conditions favouring vegetative activity, such as cutting or stress, stimulate their growth.

continued

Fig. 3.11 (a) Suckers of different ages originating from the stump of an old olive tree; (b) watersprouts; note the vigorous watersprout to the right (indicated by the arrow) overgrowing the remaining shoots; (c) vegetative shoots;

Buds can also be classified based on the organs they form upon development. Flower buds produce inflorescences, vegetative buds produce shoots, and mixed buds develop into both shoots and inflorescences. The time interval occurring between the formation and the sprouting of buds is another criterion for classification. Prompt buds develop in the same year of their formation and produce prompt lateral shoots, often called "feathers". Winter buds break during the growing season following that of their formation. Latent buds can remain inactive for years and sprout in response to manipulations or stress conditions.

Shoots are classified according to their activity (vegetative or reproductive), position, vigour, or age. Suckers are vigorous vegetative shoots originating from the root or the crown of the plant (Figs 3.2; 3.11a). Suckers tend to be numerous when the plant is ageing or it has been damaged by stress. Vegetative shoots have long internodes and do not bear flower buds. They usually grow erect on two-year-old wood (Fig. 3.11c). Watersprouts are more vigorous and have longer internodes than vegetative shoots. They grow almost vertically from latent buds on old wood (more than three years old); they are formed on the upper side of horizontal branches in inner, shaded parts of the canopy and in the tops of trees that have been headed back (Figs 3.11b, c; 5.7). Watersprouts often bear lateral shoots developing from prompt buds, and become fruitful three to four years after their emergence. Suckers and watersprouts can grow up to 2–3 m in one growing season. Both retain some of the juvenile characteristics (see also section 3.2) and should

Figure 3.12 (a) Fruiting shoot of cv. Frantoio; inflorescences developed from axillary and terminal buds; (b) mixed shoot with vegetative terminal part and inflorescences formed on axillary buds; (c) apical part of a branch bearing mixed and fruiting shoots.

be eliminated by pruning unless structural parts of the canopy need to be reformed rapidly. As an alternative to suppression, watersprouts can be headed by leaving four to six prompt lateral shoots. Watersprouts without feathers should not be headed.

Fruiting shoots proper bear only flower buds. They are thin with short internodes and located at the end of fruiting branches (Figs 3.12a; 6.8). Fruiting shoots should be eliminated by pruning since they do not produce well for more than one year. Fruiting shoots are not very common in most olive cultivars. Mixed shoots are fruiting shoots with flower buds in the axils of leaves and the vegetative terminal bud(s) (Figs 3.12b; 3.14). Mixed shoots are found primarily in the median or terminal part of the branch (Figs 3.12c; 6.8). Prompt shoots on mixed shoots often bear flower buds. Mixed shoots remain fully productive for two to four years provided there is sufficient terminal growth. They are never headed, but when terminals are less than 0.15 m in length the shoots should be eliminated (Fig. 6.9). Since mixed shoots bear most production in a vast majority of olive cultivars, they are usually referred to as fruiting shoots.

Branches are shoots at least three years old with a diameter of about 40 mm. Because of the evergreen habit of olive there are no evident signs of annual elongation on the bark and this makes classification according to age more difficult than for deciduous species (Morettini, 1972). For this reason, branches are preferably classified according to their order of insertion on the trunk or other branches (see also the box on pages 79–80 of Chapter 6). Scaffolds are the woody parts that form the permanent structure of the adult plant.

3.6 Reproductive biology and fruiting

Flowering and fruiting of olive cannot be excluded from an overview of the main biological characteristics that affect pruning techniques, but an adequate discussion of these subjects is beyond the scope of this book. Therefore, only the most relevant aspects of reproductive biology are reported here and the reader is referred to the literature for more detailed information (Lavee, 1986; Lavee, 1996; Martin et al., 1994a; Rallo, 1997; Rallo et al., 1994; Rapoport, 1997).

Loussert and Brousse (1978) identify the following phenological stages from budbreak to the beginning of fruit development: inflorescence formation, swelling of flower buds, petal differentiation, early flowering, full bloom, petal fall, fruit set, and initial fruit enlargement. In winter, flower buds are already initiated, but chilling is necessary to release them from dormancy (Rallo and Martin, 1991). Budbreak and shoot growth resume when temperatures rise above 15°C, which usually takes place in spring. Budbreak leads to the development of inflorescences, swelling of flower buds and full bloom, processes that occur sequentially, driven by prevailing temperatures.

The complete reproductive cycle takes place over two growing seasons. Flower buds are induced in the summer before flowering. Fruits are harvested 16–18 months after flower bud induction (Fig. 3.13). Flower bud differentiation occurs at the end of winter about a month before budbreak. The flowers are small, grouped in inflorescences of 15–60 (depending on the cultivar). The inflorescence is a panicle whose morphological characteristics are used for cultivar identification. At bloom it has been estimated that over 500 000 flowers are present on a single tree (Martin et al., 1994a), which indicates a remarkable reproductive effort and high demand for nutrients and carbohydrates. As a result, shoot growth is markedly influenced by the development of reproductive sinks (Rallo and Suarez, 1989). There is also an inverse relationship between the number of fruits and size of individual fruits.

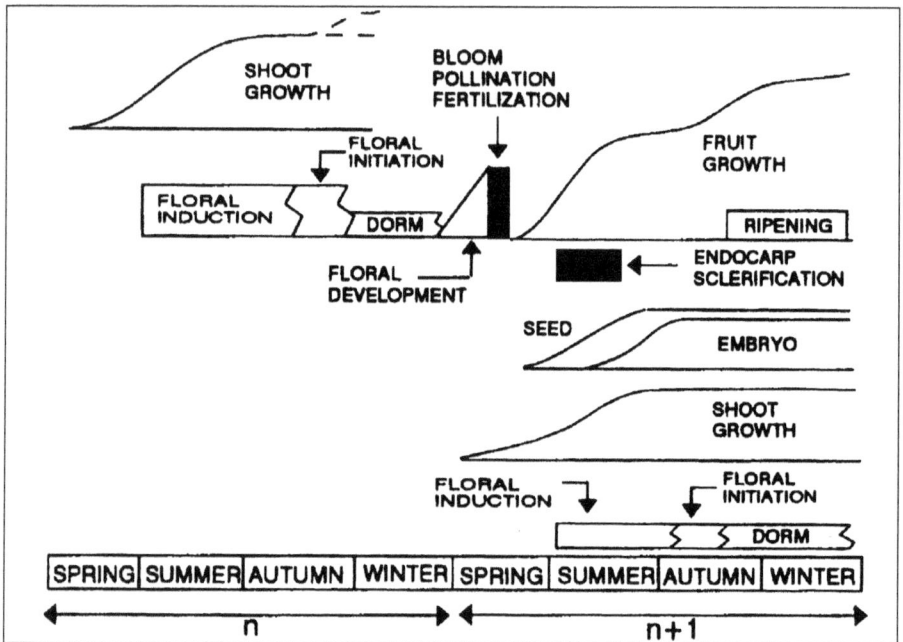

Figure 3.13 Biennial cycle of vegetative and reproductive processes in the olive (reprinted with permission from Rallo et al., 1994. *Acta Horticulturae*).

The olive flower does not have nectaries and is wind-pollinated, although in areas where flowers of other plant species are not attracting bees and other pollinators, insect pollination of olive flowers has been observed. Fruit set indicates the stage of ovary enlargement following pollination and fertilization, but is more precisely estimated about 60 days after full bloom and again before harvest to account for abscission of fruitlets. Fruit abscission usually slows down from about six weeks after full bloom, but can resume if severe stress occurs.

The olive fruit is a drupe. Fruit growth follows a double sigmoidal pattern (Fig. 3.13) and development occurs over a longer period of time than in other stonefruits (apricot, cherry, peach). Olive fruits are ready to be harvested about 180–200 days after full bloom (this interval is shorter for table cultivars which are harvested at green maturation). The different growth stages of the fruit are often not well defined (Lavee, 1986). Oil accumulation begins at the end of the pit hardening stage and increases linearly until commercial maturity, as long as environmental constraints are not limiting its biosynthesis (Lavee, 1996). Oil yield is genetically regulated, whereas oil accumulation largely depends on the environment, growing season, and biennial bearing cycle. Oil percentage on a fruit dry weight basis reaches a maximum once maturation is complete, but increases slightly on a fresh weight basis during partial dehydration of ripe fruits. Fruits at different stages of maturation are present in the canopy at the same time. This is more evident in some cultivars than in others. Ripening is almost synchronous in the cultivar Leccino whereas fruit maturation in the cultivars Frantoio and Coratina is spread over a longer period.

Heavy cropping or shading delay fruit maturation and reduce oil accumulation. Because of the high density of foliage, light levels inside the canopy can reach very low values, even below 10% of full sunlight. At this level of shading a significant decrease in flower bud initiation, differentiation, and fruit set was observed if shading was begun at an early stage (Guerriero and Vitagliano, 1973; Tombesi and Standardi, 1977). The effect of shading was cultivar-dependent. If trees were shaded from the beginning of winter, flowering was reduced in 'Frantoio' and 'Coratina', but not in 'Leccino' and 'Maurino'. Shading for only three months, between July and October (northern hemisphere), was sufficient to decrease flower bud formation significantly in 'Leccino' and 'Coratina' (Tombesi and Cartechini, 1986). Early shading (July through October) also decreased fruit size and oil yield of 'Leccino', 'Coratina' and 'Maurino', but there were no differences in these parameters if plants were shaded from December through May (Tombesi and Cartechini, 1986).

The typical fruiting shoot is the mixed shoot which consists of two distinct portions: (a) the terminal vegetative part which results from the current season's growth; (b) the one-year-old part, where flowering and fruiting occur as the result of induction and initiation processes begun in the previous season (Figs 3.13; 3.14). Since fruit set is low in olive (2–5% of the flowers) and only one to three fruits are normally set per inflorescence, yield per shoot is directly proportional to the number of inflorescences present on the fruiting shoot. Therefore, the higher the number of nodes in the terminal part of the fruiting shoot, the more inflorescences and yield potential the following year. Optimal management requires that, within each fruiting shoot, resources must be distributed so that the current year's crop is matured while vegetative growth of the distal part of the shoot remains sufficient. This balance of resources has to be maintained over a number of years to guarantee stable crops for several growing seasons.

The compromise between vegetative and reproductive activities on the same shoot has important implications for pruning. First, fruiting shoots cannot be shortened excessively or much of the current year's production will be lost. Second, because terminal growth of the fruiting shoot tends to slow down after two years of high yields, fruiting shoots and branches must be adequately renewed by pruning. The time before fruiting shoots become exhausted is longer under conditions favourable for shoot growth. Exhaustion is evident when a small cluster of leaves occurs at the end of an almost bare shoot (Fig. 6.9). As a rule of thumb, to determine the time when shoots are exhausted, one can measure the current year's growth of fruiting shoots.

Figure 3.14 Reproductive and vegetative parts of the mixed (fruiting) shoot. Fruits are present only on the one-year-old wood (n), whereas the next year's crop will be borne on the vegetative terminal of the shoot (n + 1).

If terminal growth is less than 0.2 m, the shoots are nearing exhaustion and should be renewed by pruning. Terminal growth between 0.25 and 0.60 m is adequate to maintain high yield the following year, and indicates that the shoot does not need to be pruned (indicated by n + 1 in Fig. 3.14). Terminal growth exceeding 0.6 m may be too vigorous and detrimental for stable production over the years. The intensity of pruning must be based on assessment of overall tree growth, rather than on growth of single shoots (section 4.5).

3.7 Alternate bearing

Alternate bearing or biennial bearing indicates the tendency of some tree species, including olive, to yield a heavy crop one year and little or no fruit the next year. Alternate bearing occurs at different levels. Single branches on the same tree or trees in the same orchard may be out of phase with the other branches or trees (Monselise and Goldschmidt, 1982). Such patterns of production can even become evident on a territorial scale — alternate bearing appears in regional statistics for oil production (Lavee, 1994; Morettini, 1972).

The physiological mechanisms causing alternate bearing in olive are not well understood. Several factors play a role in the regulation of this phenomenon. A heavy crop reduces the number of flowers developing on the tree the following year (Fig. 3.15). Developing fruits seem to reduce flowering via an inhibitory message released by the seed during the period of flower bud induction (Fernàndez-Escobar et al., 1992). Removal of fruits within seven weeks after full bloom increases flowering the following year (Lavee et al., 1986). If the time of harvest is postponed beyond a certain threshold date, fewer flower buds are differentiated the following year ((n + 1) in Fig.

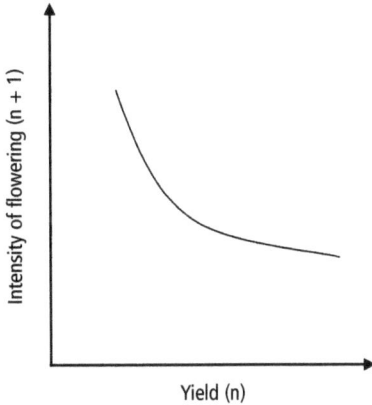

Figure 3.15 The relationship between yield in year (n) and the intensity of flowering the following year (n + 1). Not drawn to scale.

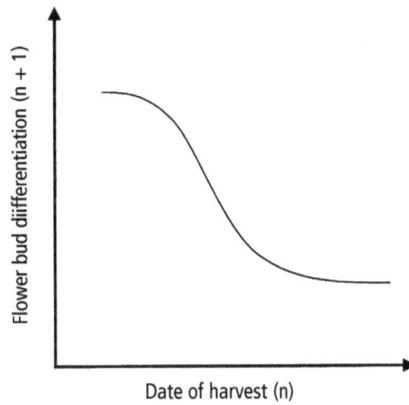

Figure 3.16 The effect of harvest date in year (n) on flower bud differentiation the following year (n + 1). Not drawn to scale.

3.16). Overlapping of vegetative and reproductive processes at some critical phenological stages, and competition for nutrients, water and carbohydrates are likely to act in conjunction to cause alternate bearing. However, the onset of alternate bearing is often triggered by a climatic event, such as spring frost at flowering or severe drought at the time of flower bud induction. The reduction in crop load is sufficient to start alternation. Alternate bearing increases with tree age, and, once established, is difficult to control. Some success has been obtained by girdling branches of some cultivars (see also section 2.9) or removing part of the crop in high cropping years. Nevertheless, neither method has commercial relevance and can be used only experimentally.

In conclusion, alternate bearing can be considered as one of the consequences of the biennial reproductive cycle. In ecological terms alternate bearing can be interpreted as an ancestral trait of the wild plant to adjust the reproductive load to external conditions.

3.8 Water use and response to abiotic stress

The olive plant can survive in areas with only 200 mm of rainfall, but needs at least 400 mm to be productive. The olive uses water more efficiently than other temperate fruit species. An estimated 312 g of water must be transpired to produce 1 g of dry matter in olive, whereas 400 and 555 g are needed in *Citrus* and *Prunus* species, respectively (Bongi and Palliotti, 1994). The same authors report that water consumption by olive is about 30 and 40% of that of *Citrus* and *Prunus* species, respectively. The high water use efficiency of olive is considered to be a result of adaptation to drought conditions. Physiological mechanisms conferring tolerance to soil water deficit include the capacity of leaves to tolerate low water potential and relative water content, osmotic adjustment, and synthesis of compatible solutes (Gucci, 1998a; 1998b; LoGullo and Salleo, 1988). The small diameter of xylem vessels causes high hydraulic resistances in the xylem pathway and makes olive plants less susceptible to cavitation during drought periods (Lo Gullo and Salleo, 1990; Salleo and Nardini, 1999).

Despite high resistance to drought, irrigation is a powerful means to regulate growth of both vegetative and reproductive organs. Irrigation requirements and cultural coefficients have been estimated for most areas where olive is grown (Goldhamer et al., 1994; Orgaz and Fereres, 1997). Irrigation is particularly advantageous during the training phase to form the permanent structure of the tree more quickly. Regulated deficit irrigation (RDI), whereby sub-optimal volumes of water are supplied to control vegetative growth and reduce competition with reproductive sinks, is used to save irrigation water (Alegre et al., 1999; Girona, 1995; Goldhamer, 1999).

Olive is sensitive to low temperature stress. Sensitivity depends on the phenological stage and the plant tissue, with flower buds and flowers being more susceptible than fruits. Shoots and old wood can tolerate a temperature of –5°C during winter without damage. However, the same temperature may be harmful if the plant is actively growing. In general, old parts (trunk, major branches) are more resistant than young shoots. There are differences between cultivars in terms of cold resistance, but most of the information available derives from empirical observations after frosts (Denney et al., 1993; Iannotta et al., 1999). Canopies damaged by cold can be regenerated by removing the damaged parts and exploiting the high potential of this species to produce new shoots from adventitious buds (section 6.9).

Olive is quite resistant to soil salinity, with cultivar differences due mainly to the capacity to exclude sodium and chlorine from the shoot and leaf tissue. Other mechanisms responsible for salt tolerance are osmotic adjustment, accumulation of compatible solutes, and drought-induced modifications. Abscission of leaves is effective in reducing the total salt load in the canopy at high levels of soil salinity. Water of low salinity may also be used to control vegetative growth and reduce tree size. Salts accumulating in the soil over the years can be reduced by proper leaching. A detailed analysis of mechanisms for salt resistance and cultural implications was recently published (Gucci and Tattini, 1997).

Chapter 4

Pruning olive trees

4.1 Objectives

There are several reasons for pruning olive trees. In young plants pruning is needed for rapid formation of a well-balanced canopy structure and early onset of production. The structure should be: (i) designed for maximum light interception and even occupation of space; (ii) strong enough to support the crop load once the plant becomes fully productive; (iii) as rigid as possible to transmit efficiently the vibration by shakers used for mechanical harvesting. In mature plants pruning is mainly required to renew the fruiting surface of the tree and achieve high yields, maintain vegetative growth of fruiting shoots, maintain the skeleton structure, contain tree size, favour light penetration and air circulation inside the canopy, permit control of pests and diseases, prevent ageing of the canopy, and eliminate dead wood.

Under certain circumstances pruning can be required to alleviate the effect of abiotic stress, to reform the canopy after damage by frost or pests, to rejuvenate old or abandoned trees, and to adapt obsolete training systems to mechanical harvesting (see also Chapter 6). In modern olive growing the training system should permit easy movement of machinery in the orchard; little attention need be paid to specific tree shapes. A complete list of the objectives of pruning olive trees is given in Table 4.1.

Table 4.1 List of objectives for pruning olive plants.

The objectives of pruning are to:
• form and maintain tree structure
• allow early onset of production
• achieve high yields
• optimize light interception by the canopy
• renew fruiting shoots and prevent ageing of the canopy
• balance vegetative and reproductive activities
• control tree size
• eliminate dead wood
• repair damage to the canopy after stress
• rejuvenate old or abandoned trees
• adapt old trees to mechanical harvesters
• facilitate pest and disease control
• improve fruit quality in table cultivars
• improve aesthetic value of plants grown as ornamentals

Not only do the objectives of pruning change during the life cycle of the orchard, but they also depend on the use of the final product. High yields and oil quality are the two main goals of cultivation for oil production. In table olives more emphasis is given to light penetration and to improving fruit quality by reducing competition among fruits. In cases when olive plants are grown as ornamentals in gardens or for landscaping purposes, pruning should be directed at achieving the maximum aesthetic value rather than yield.

4.2 Economic considerations

Pruning is second only to harvesting in cost, as it generally accounts for about 20–30% of annual cultivation costs. In a 6 x 6 m traditional orchard trained to polyconic vase at Gavorrano in southern Tuscany, pruning costs represented up to 34% of annual cultivation costs calculated over a period of eight years (Table 4.2). In this trial it took about 45 minutes to prune a single tree, but even in systems like the vasebush, where pruning took only 17 minutes per tree, the overall cost of pruning was not less than 17% of annual cultivation costs (Table 4.2). Cantini and Sillari (1998a) estimated that pruning represented 23% of the annual monetary cost of a 6 x 3 m orchard trained to a central leader monocone during four years of full production (1993–96). The monetary cost includes general expenses and taxes whereas the cultivation cost does not. Pruning costs can be reduced to 7% of the annual cost by pruning with a chain saw every two years (Table 4.2; see also section 4.6). Mechanical pruning or coppicing the orchard are the least expensive methods of pruning, but both these systems have disadvantages that restrict their application (see also sections 6.6 and 8.7).

Table 4.2 Time required for pruning and fruit yield in four olive orchards trained to either a vase or vasebush in southern Tuscany. Biennial pruning reduced costs without major effects on yield. Fruit yield is the mean of yield from 1989 through 1996. Pruning costs were estimated as the percentage of cultivation costs over the same eight-year period (modified from Cantini and Sillari, 1998a).

Site	Distance (m)	Training system	Age (years)	Frequency of pruning	Method of pruning	Time for pruning (h/ha)	Fruit (h/ha)	Pruning cost (%)
Follonica	9 x 9	Vase	100	Annual	Manual	93	4.28	20
Follonica	5 x 5	Vasebush	45	Biennial	Chain saw	27	4.68	7
Ravi	6 x 6	Vasebush	30	Annual	Manual	80	5.00	17
Gavorrano	6 x 6	Vase	18	Annual	Pneumatic tools	102	5.04	34

The high cost of pruning reflects the long time required to prune a tree and the high cost of labour. Although the unit cost of labour differs among countries, the cost of agricultural labour will continue to increase in the near future. Another problem is that skilled labour for pruning is becoming more and more scarce. Thus, the main criterion for selecting pruning techniques and training systems must be their cost, assuming that they are compatible with the biological features of the species. Current economic conditions require the development and implementation of minimum pruning strategies; systems requiring much pruning to form and maintain the tree structure are inevitably destined to be abandoned in the long run. The traditional training systems, with trees planted far apart and usually pruned to regular shapes, are old-fashioned and are gradually being replaced by free-canopy systems in more intensively cultivated groves.

The importance of the aesthetic value of certain systems should not be underrated as this criterion is always denied in theory but much appreciated in practice (see also section 7.5). A geometrical architecture is appealing to many growers but far too expensive to be achieved and maintained. Aesthetic criteria, except in gardening or landscaping, must be subordinated for practical and economic considerations. Ideally, pruning should be performed easily and quickly with unskilled labour (see also sections 7.6 and 9.3 for a discussion of economic aspects).

Pruning tools

A number of tools can be used to prune olive trees (Fig. 4.1). The pruner's kit should include pruning shears, saws, chain saw, gloves and goggles. Gloves and goggles are very useful to avoid being injured by leaves and shoots while pruning.

Pruning shears are used to cut shoots less than 20 mm in diameter. Double-bladed shears are more suitable than one-bladed shears for cutting flexible shoots. Professional pruners prefer pruning shears with shock absorbers to reduce fatigue during many hours of work. Lopping shears are not used as frequently in pruning olive as they are in pruning other fruit trees, because the evergreen foliage and the numerous thin shoots make access to the internal part of the olive canopy more difficult. They are preferred for pruning young plants rather than mature ones, and can cut branches up to 60 mm in diameter.

The hand saw is the most practical tool to cut shoots and branches ranging from 30 to 80 mm in diameter in the internal part of the canopy where the vegetation is dense. Saws can have either rigid or folding blades. The best results are obtained with a rigid blade of at least 0.4 m in length, especially for heavy work. Saws with a folding blade are relatively short and used primarily in gardens or small orchards.

The use of a chain saw is becoming more and more common to reduce the time and cost of pruning. The chain saw must be light and robust for it is virtually impossible to work with a heavy chain saw for many hours. Models weighing only 4.5 kg are suitable for pruning orchard trees. For large cuts (e.g. to major branches or the trunk) the chain saw must have at least 0.35 m of free blade to be operated efficiently. Note that using a chain saw is dangerous. This equipment should only be used by pruners in good physical condition, standing on the ground or a stable platform, wearing helmet, goggles, gloves and heavy-duty clothes for protection. Rest intervals should be frequent, and all possible precautions should be taken for the workers' safety.

Pneumatic tools, both shears and saws, can be installed on poles and can be used to prune plants up to 3.0–3.5 m in height without using ladders. Pruning heads can cut branches up to 40 mm in

continued

diameter. A tractor and a compressor are generally required for every two pruning units. Efficient operation of pneumatic tools requires two to four workers for each compressor unit to justify the initial investment. When pruning with a chain saw or pneumatic tools, adjustments in pruning methods are necessary (see also section 4.6).

All tools should be kept sharp and clean. Blade sharpening often requires professional skills, but many types of shears are sold with replaceable blades. Periodic cleaning of blades to remove wood particles can be done simultaneously with disinfection using 70% ethanol. If blades are dipped in pesticides or copper solutions during pruning to prevent spreading of diseases, they should be carefully rinsed with water and then wiped with a cloth to avoid corrosion.

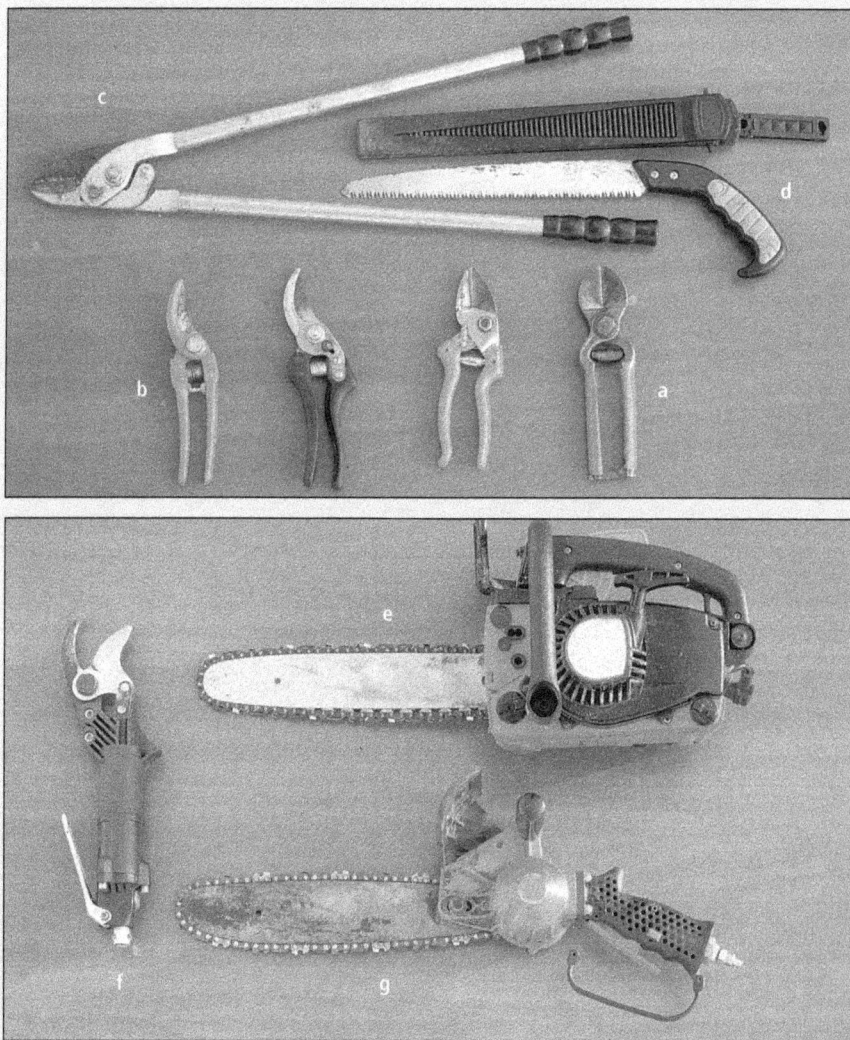

Figure 4.1 Tools and equipment commonly used for pruning olive trees. (a) double-bladed pruning shears; (b) pruning shears with shock absorber; (c) lopping shears; (d) saw; (e) chain saw; (f) pneumatic shears; (g) pneumatic saw.

4.3 Techniques

Pruning techniques vary depending on specific cultural conditions and social factors. The type of pruning must be adjusted in relation to plant age, training system, crop load, product use, environmental conditions, soil fertility, and farm structure. Given these many factors, no single method of pruning is valid for all cultural conditions. The most limiting factor, such as the availability of labour or the cost of pruning, often becomes the main criterion for choosing between alternatives (see also Chapter 7).

Despite the variability of responses to pruning, a few general rules hold under most circumstances. A list of technical guidelines for pruning according to modern criteria is summarised in Fig. 4.2. These concepts focus only on the operations strictly needed to save time and money. For example, all trees in an orchard need not be pruned every year. During training only plants without well-balanced scaffolds must be pruned regularly. Mature plants bearing large crops can also be pruned lightly or less frequently as long as renewal of fruiting shoots is adequate. Nevertheless, although olive can endure some neglect and stress, it requires the care normally used for other horticultural crops for optimum production.

1 Not all trees in a grove need to be pruned every year.

2 Adjust pruning to plant age.

3 Proceed from top to bottom.

4 Make large cuts before small ones.

5 Correct differences in vigour between branches.

6 Pruning should be rapid and simple.

7 Cost is more important than appearance.

8 All cuts that can be put off to the following year(s) should be.

Figure 4.2 Summary list of main technical guidelines for pruning trees for modern olive growing.

Pruning severity should vary with stage of growth. Young plants should not be cut severely unless they have been physically damaged and vegetative response needs to be stimulated. The intensity of pruning will increase as the plant ages, since vegetative activity tends to decline with age. The differences in techniques and severity of pruning depending on stage of development are described in detail in Chapters 5 and 6.

Pruning olive trees always means cutting shoots or branches. Inclination and bending may be occasionally employed when training plants to polyconic vase, palmette or Y-trellis, but these systems are too expensive today. Girdling, scoring, notching, and other pruning operations are not used in the field unless for research purposes (see also sections 2.7 and 2.9).

Different types of cuts can be made. The thinning cut consists in suppression of the whole shoot, or in reducing the length of the main axis by cutting close to a lateral shoot or branch, which then assumes the terminal role. The lateral shoot utilises part of the resources available in the remaining tissue to buffer the vigour of young shoots inserted near the cutting point (see also sections 2.8 and 3.5). Thinning cuts reduce branch elongation and the overall volume of the canopy, and maintain fruits and foliage closer to the centre of the plant. Figure 4.3 illustrates various examples of thinning cuts on the main axis and secondary branches of olive. Correct thinning requires that the cut be directed to oppose the natural growth habit of the plant by leaving the more erect shoots in plants with a pendulous habit (undercutting) and vice-versa (see also sections 5.6 and 6.2).

Entire shoots or branches are removed when fruiting is excessive, when there are too many competing shoots, and when sunlight cannot penetrate inside the canopy. The elimination cut reduces competition between shoots without modifying the main axis. The remaining shoots or branches are selected with different orientations to avoid mutual shading and overlapping

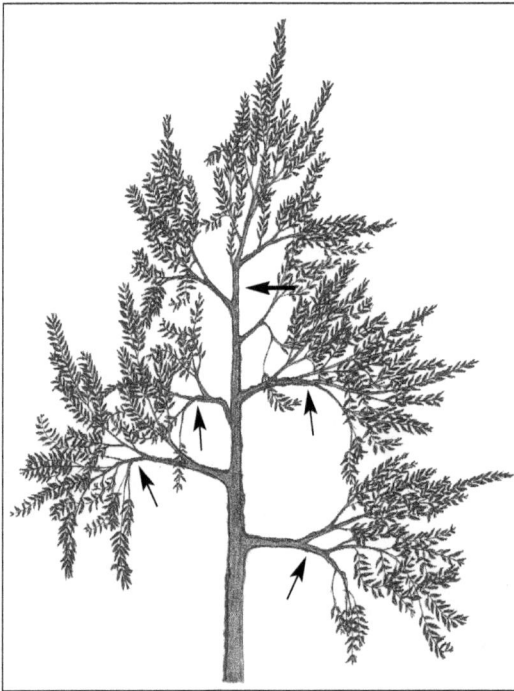

Figure 4.3 Schematic drawing showing points where heading-back cuts can be made on the main axis and secondary branches of olive. The thin arrows indicate the potential points where shoots are cut. Note the undercutting of low-hanging shoots on the lowermost secondary branches. The thick arrow indicates the removal of the main axis near a lateral shoot that then becomes the terminal.

(Fig. 4.4; Fig. 5.7), and spatially separated along the main axis (at least 50 mm if the shoots are to be the future main scaffolds; see also Chapters 5 and 8). Crowding of shoots along the main stem is frequent in both young and mature olive plants, especially in free-canopy systems where the structure is formed almost naturally with only a few cuts to correct differences in growth between scaffolds. In this case, to keep pruning costs low only few shoots are eliminated every year. Suckers, watersprouts and overcrossing shoots are typically eliminated from their point of origin on the main axis.

Figure 4.4 Elimination cut (indicated by the line) of a lateral shoot to reduce competition in a young canopy. The remaining lateral shoot (indicated by the arrow) is suitable to become a main scaffold.

The cut should be made close to the insertion point of the lateral branch, external to the branch collar and bark ridge (section 2.7). Wound repair may be delayed if the cut is made too close to the main axis or if a long stub is left. The delay in healing makes the exposed tissue a preferential point of entry and diffusion of pests and diseases. Slanting the cut avoids accumulation of rainwater, which may infiltrate down inside the bark and cause rot.

The branch collar and bark ridge are not easily identified in small shoots (Fig. 4.5). In this case one can leave a few millimetres of the stem to assure that drying out of the cut surface does not extend to the remaining axis. In all cases sharp tools should be used to make a clean cut without tearing the bark.

There is usually no need to disinfect the cut surfaces after pruning. However, if the orchard is infested with olive knot (*Pseudomonas syringae* pv. *savastanoi* E.F. Smith), foliar sprays of copper products or other low toxicity pesticides are useful to lower the probability of new infections penetrating through the cuts. Pruning tools should also be sanitized repeatedly when pruning infected trees. Foliar sprays are used a few days after hail or frost damage to prevent spreading of infections. Large cuts on old trees, trees of historical interest or those in sub-optimal growing conditions should be treated with copper products diluted in oil- or latex-based paints. Bordeaux mixture is also suitable if diluted in hot grafting wax and applied as a paste. However, the degree of protection by copper products is limited, since they are not specific for olive knot. Applications of paste or paints are time-consuming and should be done only when absolutely necessary. Large cuts should be avoided during the growing season because wound repair is more difficult then than at the beginning of spring. The wood removed by pruning can be chopped into small pieces and applied onto or tilled into the soil. Large pieces of wood should be removed from the orchard.

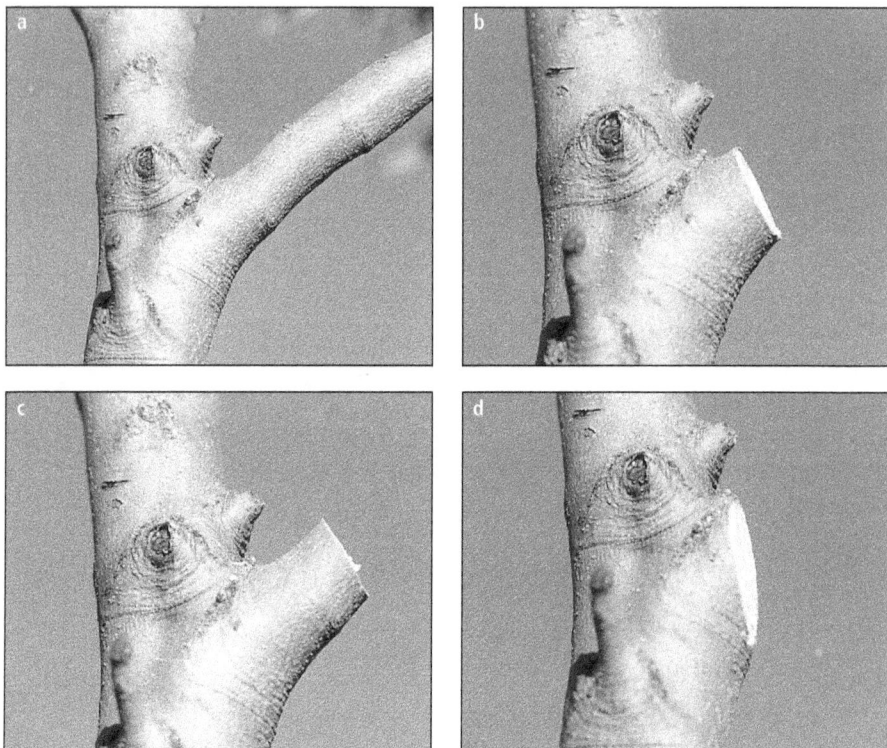

Figure 4.5 Correct procedure for removal of a lateral shoot. (a) Shoot before cutting; (b) correct cut; (c) the stub left too long after cutting; (d) cut too close to the main shoot.

Another principle to keep in mind when pruning is to proceed from the top to the bottom of the canopy to be able to assess the path of light penetration in gaps opened after each cut. The visual assessment, which is routinely done before starting pruning, can help in identifying areas of thicker foliage or differences in vigour among sectors of the canopy. This assessment requires experience, but is absolutely necessary to facilitate proper pruning.

Large cuts should be made before smaller cuts. The procedure for large cuts is to incise shoots larger than 50 mm in diameter on their lower side for about one-quarter to one-third of their thickness before the cut is completed from the upper side. Ripping the bark and/or damaging the cambium and sapwood are thereby avoided. To cut large branches the protocol requires four steps (Fig. 4.6). A shallow notch is made on the lower side of the branch about 0.2–0.3 m from the branch collar, then a second cut is made 1 m distal to the branch collar to reduce the weight of the limb. The final cut is made immediately outside the branch collar, following the same two-step procedure described for large cuts. Large branches can be cut more quickly by using a chain saw or pneumatic tools. Using this equipment is tiring and potentially dangerous, and workers must be trained and take frequent rest breaks (see also the box on pages 39–40).

Pruning is done to equalise differences in growth and size among branches and canopy sectors. This will be started from the first growing season to avoid major problems on mature trees. Heading cuts can be made to reduce the excessive vigour of some branches, thereby encouraging the growth of less vigorous ones. In sites with a strong prevailing wind the tree tends to grow flag-shaped. In this case pruning should be heavier on the downwind side, while the shoots on the upwind side should be pruned only lightly (see Figs 5.2 and 5.3).

A final recommendation is to time oneself during pruning; one can easily remove too many shoots. This is true not only for traditional methods of pruning and inexperienced labour but also for expert pruners working in intensive orchards. A good general rule is prune less than one would like in year one, leaving some shoots to be eliminated the next year. Whenever in doubt, one should prune little rather than too much.

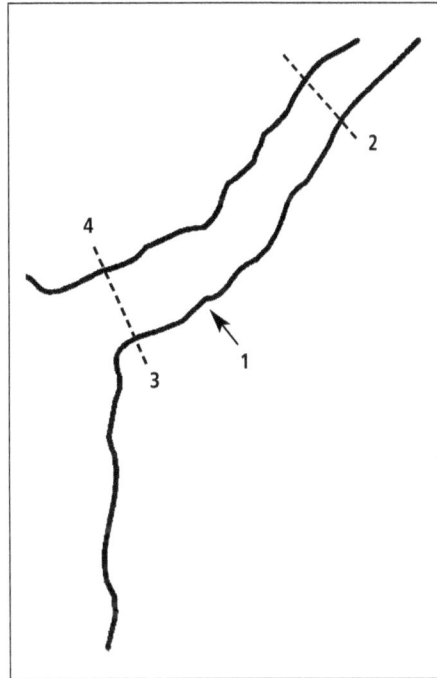

Figure 4.6 Sequence showing the proper procedure for making a large cut on a lateral branch. (1) notch; (2) first cut; (3) notch; (4) final cut.

4.4 Time of pruning

Pruning should be performed between the end of winter and flowering. Cutting stimulates metabolism and growth, which makes the plant tissue more susceptible to cold injury. In mild climates, with no spring frosts, pruning can be started in winter. Beginning pruning early may also be useful to expand the period of labour utilisation on the farm and make olive more compatible with labour needs of other crops. However, pruning before budbreak is risky in cold climates, where the high probability of frosts may damage the remaining tissues and delay wound repair. An advantage of pruning after budbreak is that even the inexperienced grower is able to assess the number of flowers and the potential crop removed by pruning, whereas flower buds cannot be distinguished macroscopically from vegetative buds at or before budbreak.

Waiting to prune until emergence of inflorescences is feasible in small orchards, but may be difficult to manage in large plantations where a longer period for pruning is necessary. Pruning should not be delayed until after full bloom, since it will remove tissues towards which nutrients and carbon reserves have already been

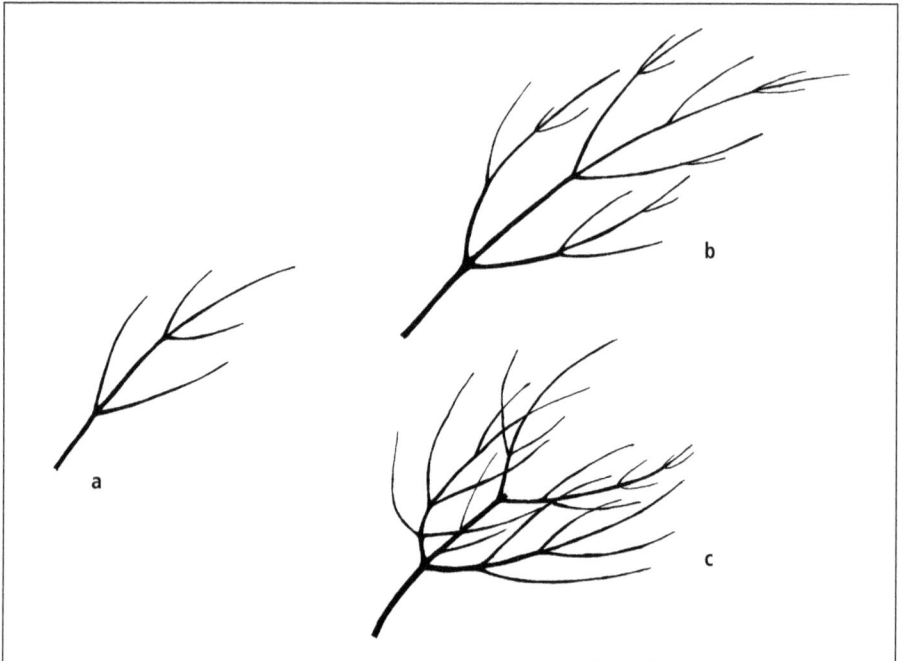

Figure 4.7 Diagrammatic representation of shoot growth if shoot (a) is allowed to grow undisturbed (b) or its apical part is removed (c).

remobilized, resulting in a net loss of resources for the plant. Late pruning does not damage the plant but can reduce seasonal vegetative growth substantially. In large plantations the timing and scheduling of pruning must be carefully planned to complete pruning within the optimal interval and allow an efficient use of labour (see also section 7.4).

Summer pruning is done during the growing season when the plant is actively growing. It is not common in cultivars used for oil and it is usually limited to the elimination of suckers and watersprouts before they become fully lignified. Their removal reduces competition for nutrients and carbohydrates during the growing season. Summer pruning is also used during the training phase to select and form the shoots that will become the permanent branches of central leader and vase systems. In cultivars used for table olives, summer pruning is used to improve fruit quality by reducing the crop load. Summer pruning can be effectively used to intervene earlier and, in conjunction with winter pruning, extend the period of pruning. However, the overall time required for both winter and summer pruning should be considered to avoid excessive pruning costs.

The timing of pruning also influences plant response. Removing shoots at budbreak results in a much more vigorous growth of the remaining shoots than if the same operation is performed at the beginning of summer. In general, the vegetative response is less when cutting is done late.

Shoot tipping (removal of the distal part of the shoot) or heading stimulates the emergence of new shoots in the remaining part, while checking growth of the main axis (Fig. 4.7). If shoot tipping is postponed until after the end of spring under non-irrigated conditions, the vegetative response is weak because of high temperatures and the development of competing sinks. In this case, shoot tipping can be used to reduce vigour of dominant branches, to re-equilibrate the canopy sectors, and to reduce vegetative activity favouring fruit growth. Summer pruning is also used to stimulate the emergence of lateral shoots at particular positions, such as in young plants to be trained to regular systems like the polyconic vase or palmette.

4.5 Intensity of pruning

The intensity of pruning should be adjusted by taking account of all those factors affecting plant vigour, including age, cultivar, crop load, soil fertility, water availability and length of growing season. As a general rule, the greater the intensity of cutting, the stronger the vegetative response of the plant (see also section 6.2). Hence, pruning should be more severe on old trees and trees of low vigour than on young plants, trees growing in irrigated conditions or in fertile soils (see also Chapters 5 and 6).

Only a few studies have focused on the effect of intensity of pruning. In the late 1960s, Guerriero and Sillari (unpublished data) applied two levels of pruning to trees trained to a vasebush, the scaffold branches of which had been previously pruned as in the polyconic vase. The removal of lateral shoots near the distal end of each of the major branches (treatment 1) was compared with no thinning of the upper lateral shoots, making only a few cuts in the more proximal part of each branch (treatment 2). Pruning took 65–90% longer in treatment 1 and the wood removed by pruning was about 0.7 kg per branch more than in treatment 2. Pruning each branch as in treatment 1 increased yield by about 0.5 kg fruit per branch over treatment 2. However, given current costs for labour and the price of olive oil in that area, it would not be economical to use the more severe pruning method. Their experiment was limited to only four years, the minimum time required to assess the effect on yield in a typically alternating crop like olive.

Hartmann et al. (1960) compared three levels of annual pruning in irrigated trees of cultivars Gordal Sevillana and Manzanilla de Sevilla planted at 9 x 9 m in California. Yields (mean of four years) were in the order: unpruned > light pruning > severe pruning. Since fruit size was not affected, Hartmann et al. (1960) concluded that under conditions of ample water availability the severity of pruning could be reduced. More recently, Garcia-Ortiz et al. (1997) showed that cumulative yield over 12 years was similar for trees trained to a three-branch vase with a single trunk and those pruned more severely in an irrigated orchard of cv. Manzanillo planted at 6 x 5 m near Cordoba. In both treatments trees were pruned every other year.

The intensity of pruning affects soil water availability. The moisture content in the 0–80 cm layer was higher in olive groves pruned severely (8000 m^3 of canopy volume per ha) than in groves pruned less severely (10 000 m³ of canopy volume per hectare) in Spain, with potential deleterious consequences for fruit growth during the period of summer drought in the less intensively pruned trees (Civantos and Pastor, 1996). Thus, more severe pruning can be used to reduce leaf area and water consumption in dryland cultivation.

The intensity of pruning should also take into account the crop load, which varies with growing season and cultural conditions. Adjusting pruning to crop load is very important in olive because of the alternate bearing habit of this crop (see sections 3.7 and 6.3). In heavy cropping years shoot growth is reduced and pruning the following year should be limited to the elimination of watersprouts and weak shoots without removing too many fruiting shoots. On the contrary, after years of low yields trees should be pruned more severely to reduce the number of new shoots and thus the yield potential.

When trees are not pruned every year, the intensity of pruning must increase with the length of the interval between prunings. The main goal is to renew the canopy by shortening the primary branches, and to remove entire secondary and tertiary branches, especially in the central part of the tree for better light penetration. In many instances light pruning can be alternated with more severe pruning every other year to maintain the balance between vegetative and reproductive activities.

The critical question is how to determine the intensity of pruning. There is no unequivocal answer, but only a number of empirical methods that the grower can use. Loussert and Brousse (1978) reported that pruning can be classified as light, moderate and severe when about 17%, 25–33% and 50% of the wood is removed, respectively. These figures obviously are merely indicative as they vary with environmental factors and cultural conditions.

One method is to assess vegetative growth after pruning. This can be done in practice by determining the number of watersprouts and the average length of one-year-old shoots in the median part of the canopy. If the current season growth on most fruiting shoots is 0.2 to 0.6 m, the level of pruning was adequate. If the current year's growth of the majority of one-year shoots is shorter than 0.15 m, pruning was too light and will have to be more intense the following spring. In the latter case, the whole canopy will have to be thinned out to stimulate growth of new shoots and to open gaps to improve light penetration. In some cases only part of the canopy shows insufficient renewal of fruiting shoots; in this case only the sector with reduced growth need be pruned more severely.

Watersprouts are numerous if above- and below-ground parts are unbalanced, if pruning and/or shading are excessive, or the plant grows too vigorously due to excessive irrigation or nitrogen fertilization. If the plant shows a high number of watersprouts on two- or three-year-old wood, emergence of suckers from the base of the trunk, and excessive growth of current-year shoots, pruning must be light. Such light pruning consists in the elimination of watersprouts or the heading of those with "feathers", leaving most of the other shoots and increasing the amount of foliage on the canopy by increasing the number of permanent secondary branches. Alternatively, the trees will not be pruned for one year (see also section 4.6).

Another indicator used to assess the plant status is the ratio between old wood (trunk and permanent branches) and young wood (one- and two-year-old shoots). Weak growth in the current season, especially if pruning was severe and cultural conditions were adequate, may have been caused by the excess of old wood on the plant, so that leaves are unable to supply the necessary energy and metabolites for maintenance of scaffolds and new growth. However, few data in the literature quantify the ratio between old and young wood. Destructive sampling of trees is not practical at the farm level, and, unlike other fruit tree species, counting shoots is practically impossible because of the evergreen foliage and large size of olive trees. A tree with too high a ratio of old to new wood is shown in Fig. 4.8. In this case, pruning should be severe to remove most of the old wood and shorten primary and secondary branches to reduce canopy volume.

Visual assessment is currently the only method that can be recommended. Visual assessment is obviously subjective. Experience can be gained by both evaluating the volume of old wood versus the foliage and measuring the vegetative growth and yield in various orchards in at least two consecutive years. An experienced grower

Figure 4.8 Forty-year-old tree showing an excess of old wood in comparison with current and one-year-old wood, resulting in a low foliage-to-old-wood ratio.

will be able to calibrate the intensity of pruning by monitoring yield, vegetative activity and wood removed in pruning over two growing seasons, whereas a less experienced person will need at least four growing seasons. Currently the tendency is to prune lightly, not only to reduce costs, but also because other effective means (e.g. irrigation, fertilisation) are available to stimulate growth and renew fruiting. As a rule of thumb, no more than 20–30% of the weight of the current canopy should be removed annually by pruning mature plants under dry cultivation, assuming there is good renewal of fruiting shoots and the canopy does not show symptoms of ageing.

4.6 Frequency of pruning

Under most circumstances olive trees are pruned each year. Annual pruning is strictly recommended when a rigid frame and a specific shape must be achieved. Annual pruning becomes indispensable in table cultivars or when shoot growth is limited by external constraints, such as low soil fertility, long summer drought, short growing season, or old age of plants. In these cases annual pruning renews the fruiting shoots and stimulates vegetative growth. Training systems with a rigid skeleton usually require annual pruning, whereas free-canopy systems can be pruned less frequently. For instance, the bush system does not need any pruning during the entire training phase (see Chapter 5 and section 8.6). Soils and climates where shoot growth is vigorous allow greater flexibility in frequency of pruning.

Less frequent pruning reduces pruning costs and the need for skilled labour in comparison with other pruning (Table 4.3). In this respect, olive for oil production is exceptional in tolerating not being pruned every year without yield losses. In addition, the current trends toward free-canopy systems do not favour annual pruning. Intervals of two, three or four years for pruning adult trees have been tried successfully in different environments.

Table 4.3 The effect of frequency of pruning on time required for hand pruning a 6 x 6 m olive orchard trained to a vasebush. Times were estimated allowing for an increase in time required for pruning individual trees when pruning was less frequent.

Frequency of pruning	Time required for pruning		
	(min/plant)	Total (hour/ha)	
		per year	per 6 years
Annual	20	92	552
Biennial	25	115	345
Triennial	30	139	278

Results from several experiments clearly confirm the advantages of less frequent pruning. Trials conducted in central Italy showed that the time needed for pruning mature trees at 6 x 6 m spacing could be reduced by 37% (calculated over a 6-year period) without appreciable effects on orchard productivity, by shifting from annual hand-pruning to a biennial interval (Table 4.3). Experiments conducted in Italy and Spain (Table 4.2; Ferreira, 1979, and Solè and Florensa, 1991, cited in Civantos and Pastor, 1996) showed that yields of trees pruned every other year were similar to those of trees pruned annually in non-irrigated orchards. In another long-term trial in two areas of Spain with sufficient annual rainfall and wide spacings, trees pruned every three or four years produced 9 and 13% more, respectively, than those pruned annually (Civantos and Pastor, 1996).

Figure 4.9 Forty-five year-old tree (cv. Frantoio) trained to vasebush. Pruning has been performed by a chain saw every other year for the last 16 years. Note the large size of cuts and the ratio between young shoots and old wood. The photograph was taken two weeks after pruning.

In a trial where biennial pruning was adopted in 1984, there were no differences in annual yields over 15 years of production in comparison with yields when pruning was performed annually (Cantini and Sillari, 1998b). The pruning technique consists in large cuts performed with a chain saw to eliminate most secondary branches and shoots on the upper side of the main branches in the internal part of the canopy. Exhausted fruiting branches are also entirely removed and primary branches are thinned, leaving a secondary branch to replace the original leader (Fig. 4.9). Then, plants are allowed to grow without any further pruning for one year, to be pruned again the following year. In this way, almost all fruiting shoots are renewed every other year. When pruning biennially the cutting is severe. By the end of the second year the canopy appears dense and disordered and needs to be pruned (Fig. 4.10).

As previously mentioned, the frequency of pruning has to be chosen based on factors such as the rate of shoot growth, crop load, training system, planting density, soil fertility and climate. The most critical factor is the rate of the current year's shoot growth. If active shoot growth is maintained, pruning can be postponed until the following year. Pruning every two years or longer can be more easily implemented in irrigated orchards, in fertile soils, and with trees planted at wide spacing. The maximum interval between two successive prunings must be calculated by evaluating yields over the whole time interval and the vigour of trees at the time of the second pruning.

A biennial frequency of pruning can be adopted in the majority of cultural conditions, but intervals longer than three or four years are not always suitable. Implementing a triennial or quadrennial pruning should be done only if soil fertility, water availability, and climatic conditions are favourable. Otherwise, yields decline markedly, and at the end of the four-year cycle, pruning will have to

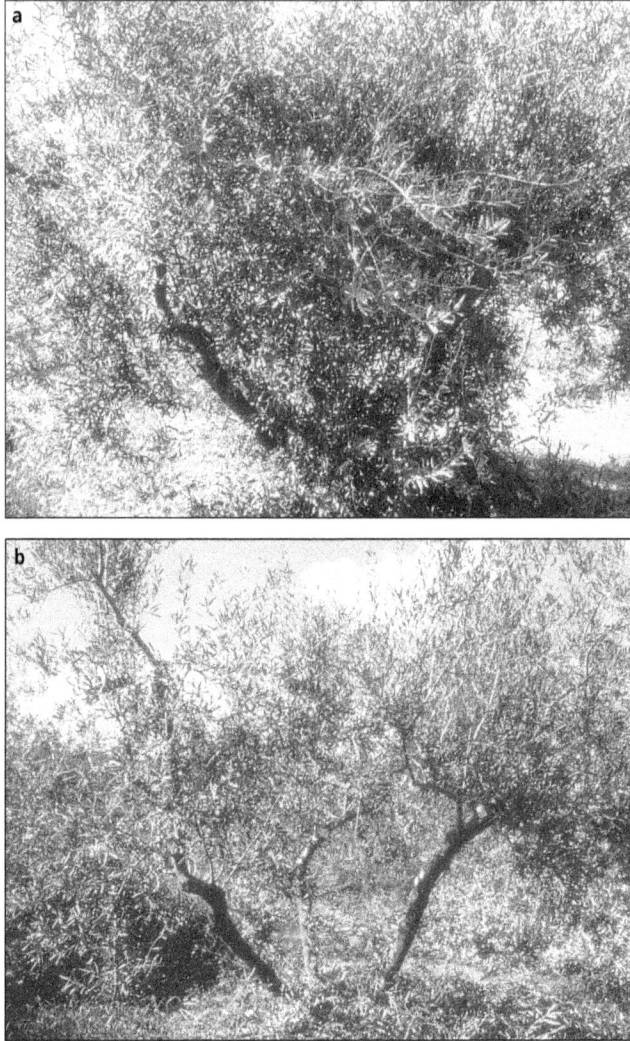

Figure 4.10 Severe biennial pruning with a chain saw of an olive tree trained to vasebush: (a) before pruning; (b) after pruning.

be drastic, with consequences on the vegetative-reproductive balance of the plant. Cultivars with an upright habit and those sensitive to foliage disease are less suitable for infrequent pruning because the excessively thick canopy and upright growth will make harvesting and pest control more difficult and time-consuming.

An interesting application of reduced frequency of pruning has been developed in Spain to rejuvenate old or abandoned trees (Garcia-Ortiz et al., 1997; Pastor, 1989). Periodical cuts are made with a chain saw only, each aimed at removing one of the two to three primary branches on the main trunk of the tree. Different branches are shortened alternately. The pruned branch, cut a few centimetres above the insertion point on the trunk, is allowed to grow freely until the onset of production (which occurs after two to four years), while another branch is cut the following year. In this way, a crop is produced on the remaining branches every year, while the canopy is rejuvenated, the size is reduced and a high leaf-to-old wood ratio is obtained. The vegetative response to the cutting of only one or two branches of the canopy is less than if all the canopy had been removed at once (section 6.9).

4.7 Strategies

It is already apparent from the above sections that there is no single method for pruning olive trees. Techniques, timing, intensity and frequency can be combined to develop strategies that suit the specific characteristics of trees and the organizational needs of the farm. To develop pruning strategies, all possible methods to control and manipulate plant growth and fruiting should be used for periods longer than one growing season.

When designing pruning strategies, one should be able to evaluate the current vigour, productivity and potential yield of the orchard. Adjustments in pruning may be necessary from site to site. Assessment of tree vigour and productivity, as well as practices like fertilisation, weed and pest control, should be performed as carefully as in annually pruned orchards. The compatibility of techniques and strategies with the long-term performance of the plant must be considered as well. An example of this is the use of mechanical pruning over many consecutive years, which causes the development of a thick, unproductive hedge on the outer shell of the canopy unless it is combined with manual pruning (section 6.6).

The current tendency is to prune olive as little as possible. Concepts of minimum pruning should be applied in all possible cases to reduce costs substantially and simplify pruning management. Minimum pruning concepts, which are further discussed in the chapters that follow, can be summarized as:

- prune only the trees that need it;
- reduce the frequency of pruning;
- adopt free-canopy systems;
- use irrigation and fertilisation to stimulate growth and sustain fruiting.

However, minimum pruning does not mean neglectful pruning. The olive tree should not be left unpruned for longer than 8–10 years under optimal climatic and cultural conditions (see also Chapters 5 and 8), but the interval between two successive prunings should not exceed three or four years under most circumstances. Intervals longer than four years without pruning result in ageing of the fruiting surface of the canopy, with consequent reductions in yield.

Strategies of minimum pruning should be implemented (e.g. long intervals between pruning, trees unpruned during the training phase) only if the pruner has a clear picture of the physiological status of the tree. If he/she cannot interpret the health, vigour and productivity, it is probably better to prune annually rather than less frequently. Minimal pruning will help to reduce costs also when trees are pruned annually; reducing the number of shoots cut per year saves pruning time.

The final criterion in deciding on the most suitable pruning strategy is economics. Every time the frequency or the intensity of pruning is decreased, the cost is reduced. Although annual pruning is still the most common method in olive-growing areas, it is expensive and in many cases it removes too much foliage with no increase in yield or oil quality. Increasing the time between two successive prunings is a simple and effective way to reduce costs in the olive grove. In practice, the grower can first try annual pruning until he/she becomes familiar with plant response to type and severity of cutting, and then change to biennial pruning. This simple shift often reduces costs by half. If problems are encountered in the biennial system one can easily revert to annual pruning, although our experience is that pruning intervals longer than one year allow more flexibility in orchard operations. Note that a biennial pruning regime does not imply a doubling of the time needed to prune a single tree. This is because fewer but larger cuts are made and the time needed by operations such as positioning of ladders, transport of equipment, or movement of machinery does not change with the number of cuts made on each individual plant (see also section 7.4).

Chapter 5

Pruning young trees

5.1 Objectives

The training period starts at planting and ends once the formation of the permanent structure is completed and the plant reaches full production. The objectives of pruning during the training phase are to allow fast growth of the plant, to form a well-balanced canopy, and to bring trees into production early. The structure should be designed for maximum light interception and occupation of space and should be strong enough to support heavy cropping once the plant becomes fully productive. For trees to be mechanically harvested by trunk shakers, pruning should be aimed at forming a single trunk for easy attachment of the mechanical clamp and a structure as rigid as possible for efficient transmission of the vibration.

All these objectives can be fulfilled by a few pruning operations, and sometimes no pruning at all, in the first few years after planting. Nevertheless, errors in pruning made during the training period may be carried over for years and result in lower revenues. For example, severe cutting will prolong the unproductive phase of young plants, whereas the effect of pruning on the onset of production is irrelevant if plants are well managed. Plants usually come into bearing in the third or fourth year after planting.

5.2 Selecting the plant material

The quality of the plant material must be a primary concern when buying plants from the nursery. Plants well prepared in the nursery are convenient, even if more expensive, since they start producing early and require less pruning care than poor quality plants. Buying cheap, small plants is often paid back by higher pruning costs during training in the field or by delayed onset of production. Certified plant material that guarantees the genetic origin and the absence of olive knot, Verticillium wilt, and viruses is preferred.

Qualitative characteristics to be checked for are size and health of the entire plant including the root system, and presence, vigour and distribution of laterals along the main stem. Material suitable for planting should be uniform, with green foliage and without signs of juvenility, diseases or parasites. Major pests to be checked for are the moths (*Prays oleae* Bern. and *Palpita unionalis* HB), scales (in particular the black scale *Saissetia oleae* Olivier), the weevil (*Otiorrhynchus* spp.), and *Eriophides* mites. The three most dangerous diseases are the peacock's spot (*Spilocea oleaginea* Castagne), the Verticillium wilt (*Verticillium dahliae*), and the olive knot. Symptoms of sooty mold (*Capnodium* spp., *Alternaria* spp.) on leaves are indicative of plants neglected in the nursery (shading, poor control of scales).

Propagation and plant production

Several techniques can be used to propagate olive plants in the nursery. About 70% of plants produced worldwide are from rooted cuttings. Cuttings are preferably taken from one-year-old wood, well exposed to sunlight and in good nutritional conditions. Cuttings develop sufficient roots to withstand transplanting after 50–70 days (depending on the sampling period) under mist with basal heating in a glasshouse. Grafting is mainly used to propagate cultivars that are hard to root, or when the material on mother plants is scarce. Both methods produce excellent plants, but grafted plants are sometimes less uniform than rooted cuttings. The main advantage of plants produced from rooted cuttings is that, in the case of severe damage in the field, the canopy can be eliminated without any risk that new suckers will originate from the rootstock.

The main root of an olive seedling grows downward and develops into a taproot. The taproot anchors the plant in the substrate during the first period after germination. The root system of grafted plants, as well as that of rooted cuttings, also lacks the taproot, which is cut off in the nursery at transplanting. The root system is mainly formed by lateral roots originating from the crown of the plant and expanding rather superficially, regardless of the propagation technique. The lateral roots of grafted plants are usually more spaced along the main axis than those of rooted cuttings, which develop at first almost exclusively from the base of the cutting. These small differences in the morphology of the young root tend to disappear with age. The root systems of mature plants are similar regardless of the propagation technique. There is no experimental evidence showing that plants produced by either rooted cutting or grafting have better anchorage or drought resistance. Guerriero et al. (1972) compared grafted plants *versus* rooted cuttings of five cultivars in two different trials in Tuscany and Sicily and found no consistent differences in productivity or plant vigour within six years from planting. Plants should be planted in the field at about the same depth as in the nursery or container. The graft union should be about 50 mm below the soil level to facilitate the emergence of roots from the scion.

Plants should be planted in the field at about the same depth as in the nursery or container. The graft union should be about 50 mm below the soil level to facilitate the emergence of roots from the scion. Plants are sold either in containers or bare rooted. Container-grown plants guarantee the higher percentage of survival after transplanting in the field. Bare-rooted plants need more care since their roots must be kept moist at all times. Nurseries sell plants from one to four years old. It is better not to buy plants older than two years as they cost more, are more susceptible to transplant crisis, and their root system has been constrained in the container for too long. Moreover, the canopy of young plants can be trained to the final shape more easily. At purchasing it is useful to check the size of the stem and plant height, and to inspect foliage, roots and graft union for parasites and diseases (see also section 5.2). Plants suitable for planting in the field should have a well-developed and expanded root system with visible root tips and hairs in all directions, a distinct vertical stem, a number of lateral shoots evenly spaced along the main axis and possibly inserted at a distance from the crown of between 0.6 and 1.0 m. The vertical axis should be at least 0.7 m high with a diameter of at least 6 mm at 0.1 m above the crown. Plants with a height between 0.7 and 1.2 m are excellent for planting in the field.

The root system should be well expanded, healthy (exempt from root rot caused by *Armillaria mellea* Vahl. Quel.), not too dense and tortuous which may indicate ageing and too long a time in the container. The graft union should be robust and free of any pest or disease. The vertical axis should be at least 0.7 m high (minimum distance between the apex and the crown) with a diameter of at least 6 mm at 0.1 m above the crown. Plants between 0.7 and 1.2 m in height are excellent for planting in the field and can be produced in 12–18 months, depending on the propagation technique. Grafted plants usually take more time to be produced than rooted cuttings.

If plants smaller than the above dimensions are planted to be trained to single-trunk systems, it is best to favour the growth of the main axis during the first two years by firmly securing the stem to a vertical trainer and eliminating vigorous laterals. If the removal of vigorous laterals is delayed, their subsequent elimination will check growth of the whole plant. This is especially so for those in the lowermost part of the main axis, which tend to compete strongly with the vertical stem (see section 4.3). Relatively small plants can be used for free-canopy systems without a single trunk. Experimental trials conducted in central Italy have shown that plants of either 0.5 or 1.2 m in height develop a similar canopy volume after six years from planting if grown unpruned. However, it should be kept in mind that small plants are subject to greater damage by pests or cold. The *Palpita unionalis* moth and the weevil are particularly dangerous during the training phase because they can defoliate the canopy in a few weeks. Particular care should also be devoted to the elimination of weeds which compete actively for water, nutrients and light during the first three years after planting. For these reasons, it is better to buy plants of at least 0.7 m in height.

In some olive-growing areas, it has become quite customary lately that nurseries offer three- or four-year-old plants up to 1.5–2.0 m in height for ready-effect orchards. Apart from the impressive appearance at planting, there are no real advantages in buying plants of this size. These plants are generally expensive, with a root system constrained in a container for too long, and more difficult to adapt to specific training systems. In addition, large plants acclimatize less easily than small ones to transplanting, which may cause problems in arid areas without irrigation or wind-swept areas. Plants with lateral shoots originating at about 1.5 m from the crown can only be used to form traditional vase systems, unless a lot of pruning is done in the first few years after planting to adapt them to other training systems. Plants higher than 1.5 m or older than two years at planting are generally less suitable for most training systems as they need to be pruned heavily to stimulate new shoots in the lower part of the axis. When most of the plant canopy is eliminated, it is likely that one or two years will be wasted in the field to form new structures.

Not all plants in the nursery are equally suitable for different training systems (see also section 5.5). In most cases plants only need to be pruned lightly (one or two cuts) at planting and then allowed to grow without any further pruning for two to six years depending on their vigour. On the other hand, plants must be pruned annually during the training period to obtain more regular systems such as the central leader, the polyconic vase or Y-trellis.

An example of nursery plants with a different structure is shown in Fig. 5.1. Both plants (cv. Leccino) are 18 months old and about 1.2 m high. The plant on the left has lateral shoots at about 0.7 m from the top of the container and it is well suited to form trees with a single trunk. This plant will not need any pruning for the first two years after planting if the canopy is allowed to grow freely. A few lateral shoots close to the apex of the main stem will have to be eliminated for central leader systems (see also section 5.3). Laterals originating at 0.7 m above the crown can also be used as major branches of vase systems. The plant on the right of

Figure 5.1 Eighteen-month-old olive plants (cv. Leccino) ready for planting. Note the different distribution of shoots in the basal part of the main stem of the two plants. The plant on the left is more suitable for central leader or single-trunk systems, the one on the right for a vasebush or bush.

Fig. 5.1 has well-developed lateral shoots in the lower part of the main stem, between 0.2 and 0.4 m from the top of the container. This plant can be trained easily to form a vase with a short trunk, or a bush (see section 8.1). In the former case the main stem will be headed at about 0.7 m in height at the end of the second or third growing season. Note that if the growth of lateral shoots relative to that of the main stem is vigorous, there may be no need to head the stem because it will be naturally suppressed as the plant grows. No pruning will be needed to obtain a bush (section 8.6).

5.3 General concepts

The concept of reducing pruning to a minimum in modern olive growing should be applied starting from the training phase. Both obvious economic considerations and biological reasons support the statement that very little pruning has to be done during training.

Experimental evidence on the convenience of reduced pruning during training has been produced by several authors. Table 5.1 reports the fruit yield per plant cumulated over the first five years after the onset of production in four different trials in central Italy. In each trial, training systems requiring little or no pruning (bush, free vase) were compared with others which needed more pruning during the training phase (central leader, Y-trellis). The vase produced more than other systems at the Rieti and Grosseto (dryland) locations. Yield in the first few years was significantly higher in trees which had received little or no pruning at all (vase, bush) in the trials at Grosseto and Rosignano (Table 5.1). In no case did the more pruning-intensive systems (Y-trellis, central leader) yield more or produce earlier than the plants trained to free-canopy systems.

Table 5.1 Productivity of olive plantings trained to different systems during the first five years of production.

Site	Distance (m)	Training system	Year of planting	Cultivation	Fruit yield (kg/plant)
Grosseto[1]	6 x 6	Vase	1986	Irrigated	42.9
		Bush			43.2
		Central leader			23.6
Grosseto[1]	6 x 6	Vase	1986	Dryland	41.8
		Bush			31.3
		Central leader			16.1
Perugia[2]	5 x 5	Vase	1986	Irrigated	21.5
		Central leader			19.3
Rieti[3]	5 x 5	Vase	1986	Dryland	90.5
		Central leader			77.0
		Y-trellis			71.4
Rosignano[4]	6 x 6	Bush	1990	Irrigated	20.7
		Central leader			17.8

[1] ARSIA, 1994;
[2] Proietti et al., 1998;
[3] Parlati et al., 1996;
[4] ARSIA, 1999.

These data confirm those reported by Hartmann et al. (1960), who compared two methods of pruning during the training phase in an irrigated orchard of cv. Mission in California. Plants were either pruned, starting from the first year after planting, to form a vase with three to five branches, or left unpruned until the onset of production. In the first three years of production, yield of pruned trees was only 74% of that of unpruned trees, but differences in yield disappeared after six years of production. Production started in the same year for both treatments.

These results show unequivocally that unpruned plants do not yield less or later than plants pruned during the training phase. One of the reasons why unpruned plants grow more rapidly than pruned ones is that the removal of too many leaves and shoots often reduces drastically the availability of assimilates in young plants with small photosynthetic area. Moreover, severe pruning of young plants also decreases the shoot/root ratio considerably and induces a strong vegetative response that delays the onset of production. It can then be safe to recommend pruning plants as little as possible in the first five years after planting.

However, there can be specific reasons to prune during the training phase. Reasons for pruning can be summarized as follows (see section 5.4 for technical details): (a) to eliminate watersprouts and suckers; (b) to balance the development of the different canopy sectors; (c) to avoid overcrossing of shoots; (d) to reduce competition between lateral shoots and the main axis; (e) to stimulate emergence of lateral shoots at a particular point on the main stem; (f) to remove lateral shoots from the basal part of the trunk in single-trunk systems.

Although it is impossible to encompass all possible situations occurring in the field, the following concepts can be used as general guidelines:

1 the more pruning during training, the later the onset of production and the lower the yield in the first years;

2 the more regular the skeleton structure to be achieved at maturity, the more frequent and intense the pruning should be in the first few years after planting;

3 the only shoots to be eliminated are those that may compromise the definitive shape of the plant (watersprouts and suckers in particular);

4 the lateral shoots competing vigorously with the main stem are to be eliminated (arrows in Fig. 5.7) to promote the growth of the central leader or of primary branches. Shoots inserted at the same height compete strongly with one another and the main axis and will seldom develop into strong scaffolds. These shoots must be thinned to avoid their progressive weakening;

5 pruning should be gradual during the training phase.

5.4 Techniques

As mentioned in the previous paragraph, there can be specific reasons to prune young plants. During the training phase it is better to eliminate entire shoots rather than heading them. The thinning cut is preferably done next to a lateral shoot. By repeated thinning, it is also possible to widen the angle of inclination of the future lateral scaffold. Thinning cuts are always used to eliminate vigorous shoots such as watersprouts or suckers.

Heading cuts are to be used with great care during training as they tend to stimulate the emergence of many vigorous shoots next to the cut surface. Further pruning will be required in the growing season following the cut to thin the population of vigorous shoots and watersprouts. Heading the main axis of young plants is used to stimulate the emergence and growth of lateral shoots at a particular point on the axis, or strengthen the vigour of lateral shoots that are too weak to assume the role of major branches in vase systems. This operation is particularly important if plants to be trained to vase systems have not been specifically prepared in the nursery. Similarly, lateral shoots can be headed above a bud to stimulate growth of secondary or tertiary branches.

Lateral scaffolds should be selected with a vertical separation of at least 50 mm to reduce competition between them. Optimal spacing between major branches capable of supporting heavy crops is 0.1–0.2 m. Lateral branches should be oriented in different directions to avoid overlapping and mutual shading. In the polyconic vase or central leader systems they are preferably chosen in elicoidal arrangement around the main axis (section 8.4; Figs 1.3; 1.4; 8.9). Long lateral shoots overcrossing from one side to the opposite of the canopy are usually eliminated. This operation is done gradually over the years by suppressing few overcrossing shoots every year.

Removal of watersprouts and suckers is usually done in the summer, beginning from the first growing season. Alternatively watersprouts can be weakened by bending or twisting when still flexible and not fully elongated. In doing so the vigour of watersprouts is reduced and they assume the features typical of fruiting shoots within the next two growing seasons. A high number of watersprouts or suckers in young plants may be a symptom of inadequate orchard management (see also section 4.5).

Dominant shoots or branches that tend to overgrow others should also be thinned. Vigorous shoots may develop naturally in the canopy, beginning from the first growing season and usually becoming dominant in the subsequent years. Figure 5.2 shows an example of an unbalanced tree where the left side is stronger and the foliage denser than the right side. The branch indicated by the arrow is far more vigorous that the others and should have been eliminated or weakened by a heading cut at planting. At this stage there are two main options available to bring the tree back to an acceptable structure: (a) to remove the entire branch, or (b) to head back the branch itself (cut no. 1) and eliminate several shoots on that branch to reduce its vigour. In the former case there will be a substantial loss of leaf area and wood; in the latter the pruning will be more time-consuming and expensive than if done at planting. The elimination of a watersprout (cut no. 2) and the undercutting (cut no. 3) of the dominant branch are also shown (Fig. 5.2).

An unbalanced growth of branches or part of the canopy is frequent in areas with a strong prevailing wind. In such cases the canopy assumes a flag-shaped appearance (Fig. 5.3) and should be pruned so that the shoots and branches on the upwind side are

Figure 5.2 Young olive tree with the left side more vigorous than the right side. The branch indicated by the arrow is dominant over the others and should have been eliminated or weakened earlier on. The numbers indicate the sequence of pruning cuts needed to start balancing the canopy: (1) elimination of most of the dominant branch; (2) elimination of watersprout; (3) undercutting.

favoured against those on the downwind part which will be appropriately thinned. In the specific case of Fig. 5.3 the main remedy is to secure the tree tightly to a strong trainer post (of at least 50 mm in diameter and 1.5 m above ground level) for mechanical support of the trunk which is too thin for the height of the tree. The trainer post should be positioned upwind from the tree so that the bark is not damaged by brushing or hitting against it. Olive plants with a high height/width ratio are more subject to the action of wind than systems with a more expanded canopy (vase, vasebush, bush).

Figure 5.3 Flag-shaped young olive tree growing in a wind-swept area. The absence of a strong trainer post, and the low width/height ratio of the canopy make the thin trunk unable to withstand the action of wind. Too much leaf area was removed from the basal part of the main stem in the first growing season.

In conclusion, pruning during training must be as light as possible, allowing the plant to grow according to its natural growth habit and to produce early and abundantly. Under normal conditions of vigour the only annual pruning will be to eliminate the shoots like watersprouts and suckers that could compromise the final structure of the tree, and few overcrossing shoots.

5.5 Different training systems

The choice of the training system strongly affects the extent and type of pruning during the training phase. However, the variability in plant material, climatic factors and soil fertility may require adjustments in pruning even for different trees within the same orchard.

The flow chart represented in Fig. 5.4 illustrates the options available for pruning during training. Of course, the training system must have been selected long before one starts pruning the young plant. A synthetic list of the prunings during the first five years after planting is also reported in Fig. 5.4 (for details on the different training systems see Chapter 8). These guidelines apply for a container-grown plant of about two years and 1.2 m in height with evenly distributed lateral shoots along the central axis.

The first question is whether a single trunk is needed or not. The single trunk is indispensable if trees will be harvested mechanically by trunk shakers, whereas there are virtually no limitations in the choice of the training system if harvesting is

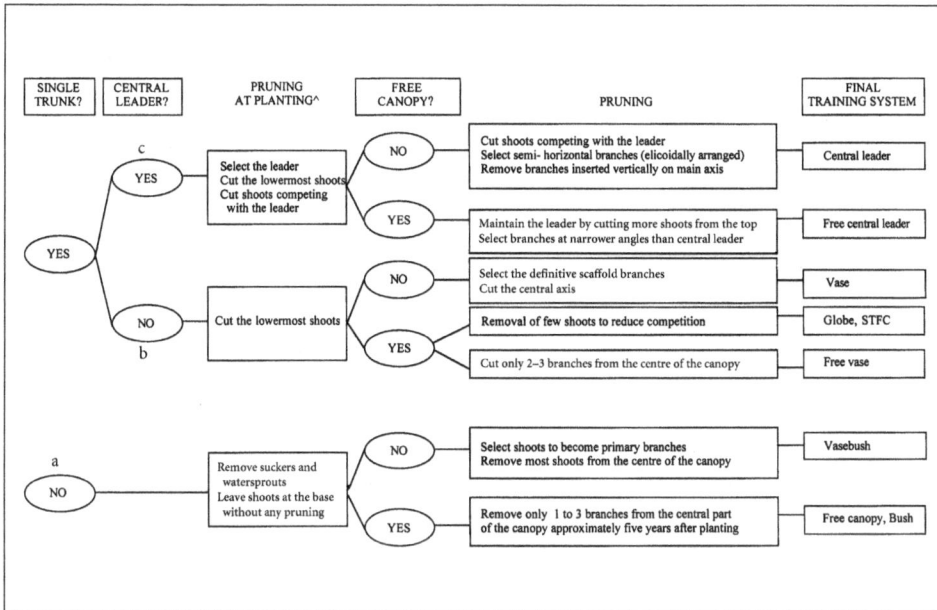

Figure 5.4 Diagram showing the options available for pruning young plants, depending on the final training system. The main pruning operations during the training phase are also reported. The sequence of steps is from left to right.

^The pruning at planting includes summer pruning during the first growing season.

done by hand or hand-operated equipment (see also section 7.2). To obtain a single trunk the young plant has to be tied to a vertical trainer at planting and lateral shoots present on the basal part of the main axis (1 m from the ground) must be eliminated. To avoid too drastic a reduction in photosynthetic area and a checking of plant growth and girth increment, the suppression of these lateral shoots should not be completed at planting or within the first growing season, but should be done gradually over the first five years after planting. Vigorous shoots or excessively low-hanging shoots are eliminated before weak shoots. In a well-balanced plant no more than three shoots will be eliminated from this part of the stem every year. If vigorous shoots are removed late, wound repair takes more time and the resulting scar may represent a point of discontinuity for attachment of the mechanical clamp, with greater risks of bark damage.

The second question deals with the presence or not of a central leader (Fig. 5.4). At the moment there are no good technical reasons to justify the adoption of the central leader *versus* other modern training systems (see also Chapters 7, 8 and 9). The maintenance of a central leader in a naturally basitonic species like olive requires regular annual pruning, which can become quite expensive over the years. The decision whether to use a central leader or not is left to the grower's personal preference. To obtain a central leader the arrow of the main axis must be kept free of lateral shoots that would compete for light and nutrients. It is usually sufficient to remove laterals originating from the more distal part of the main axis (0.2 to 0.4 m according to the size of the plant; Fig. 5.5).

This pruning needs to be done every year for the first three years and often it has to be continued after the onset of production. In systems without a central leader, there is no need to eliminate most lateral shoots competing with the main stem (Fig. 5.4).

The third option looks at whether to aim for a free-canopy system or not. There is currently a tendency toward minimum pruning. The concept of minimum pruning stems from the need to keep pruning costs low while seconding the natural growth habit and reproductive biology of the species. Free-canopy systems usually require less pruning than more regularly shaped systems (vase, central leader), which in turn are less

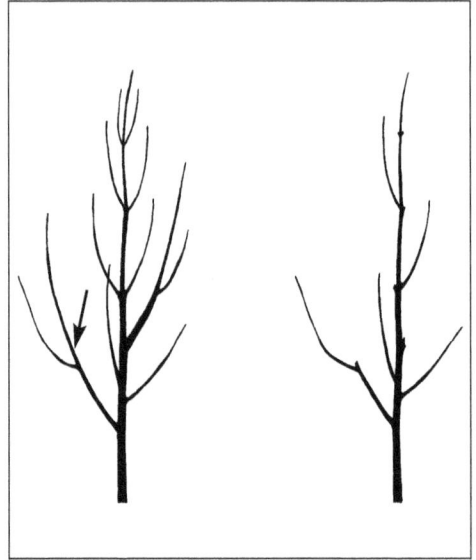

Figure 5.5 Thinning of shoots along the main axis to strengthen the central leader and lateral shoots that will form the scaffold branches of the tree. Note the heading-back cut (indicated by arrow) of the lowermost lateral shoot to obtain a wider angle of inclination.

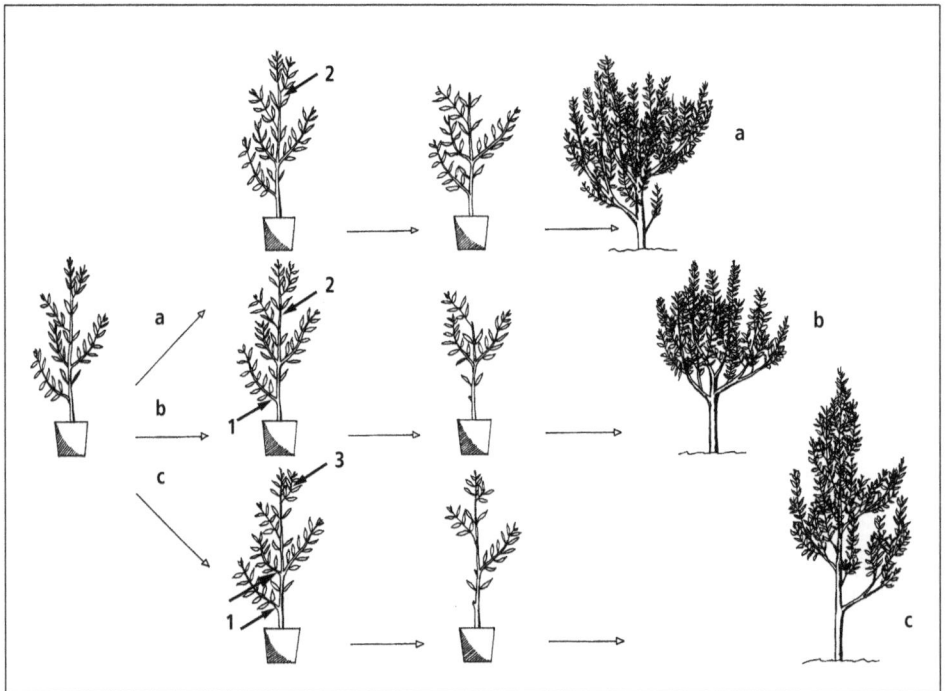

Figure 5.6 Schematic diagram of pruning young plants. Three paths are represented leading to the following final training systems: (a) vasebush or bush; (b) vase; single-trunk free-canopy; (c) central leader. Cuts 1, 2, 3 indicate removal of shoot, heading of the main axis, and heading (tipping) of a lateral shoot, respectively. Not drawn to scale.

constrained than the traditional vase systems or the polyconic vase. Free-canopy systems allow the grower to save many hours of skilled work per hectare every year. Nevertheless, the inexperienced grower often feels more at ease with a regular shape, which can be obtained by applying instructions for pruning more rigidly.

Figure 5.6 shows schematically the pruning operations required to obtain a single trunk and/or a central leader in the first year after planting. The three paths correspond to the three main routes of Fig. 5.4 and are indicated by similar letters.

To train the tree to a central leader one or more shoots needs to be cut from the lower half of the main axis to prevent them becoming as vigorous or more vigorous than the central leader of the young plant (Fig. 5.4; Fig. 5.6a, cuts no. 1). One should not leave too many shoots inserted at the same height; only one lateral shoot per node is usually left (Fig. 5.5). To assess the relative vigour of the central leader with competing lateral shoots, one can visually compare the diameter of the main stem below and above the point of insertion of lateral shoots. If the main axis is thin, there is a high probability that the central leader will be overgrown by the lateral shoots in the following years (Fig. 5.7). A third useful pruning operation is to head back (cut no. 3) or to eliminate shoots growing near the apex of the main stem. This is usually done for central leader trees by summer pruning because it is easier to assess plant vigour and select the shoots to be eliminated. In the following two years it will be important to continue to favour the central leader by pruning the lateral shoots next to it.

In the free central leader or single-trunk free-canopy (STFC) systems, pruning is less schematic and branches are selected more freely than in the central leader proper

Figure 5.7 Excessive crowding of lateral shoots that are prevailing over the central stem. This competition prevents the achievement of a vigorous central leader. The arrows point to the difference in size above and below the insertion of lateral shoots. Note the watersprout to the right of the upper arrow.

(section 8.5). Branches are chosen among the naturally developing shoots for best light interception and penetration in the lower part of the canopy. In systems with a single trunk but trained to a free canopy rather than to central leader (e.g. vase, globe, STFC) the first pruning on plants from the nursery is aimed at eliminating vigorous shoots growing in the lowermost part of the stem as shown in the previous example (cut no. 1 in Fig. 5.6b). Heading the central axis (cut no. 2) is not strictly needed, but becomes necessary to stimulate the emergence and growth of new shoots from below the cut. If good lateral shoots are already present the suppression of the central leader to form an open canopy can be postponed.

The first pruning cut on a plant to be trained to vase with a short trunk is done in the third year after planting (Fig. 5.8), if the plant had been prepared as in Fig. 5.6. The goal of pruning is to improve light penetration inside the canopy, allowing more space to the branches that will form the permanent structure of the canopy. Then, the plant can be grown without further pruning for about two to four years to form a free vase (Figs 5.8b; 8.15). The same plant can be pruned more rigidly to

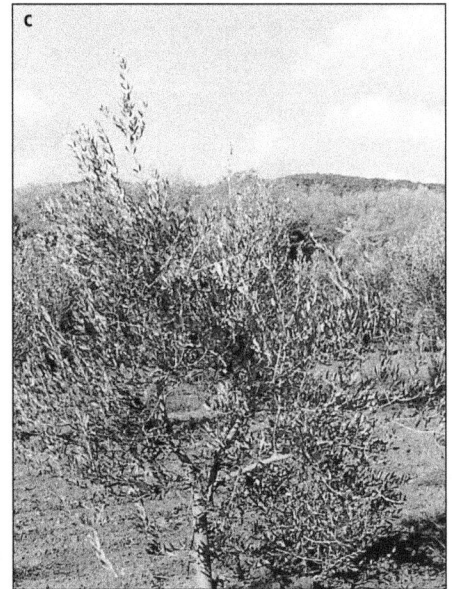

Figure 5.8 Olive plant (cv. Leccino) in the third growing season after planting. (a) before pruning; (b) after moderate pruning to be trained to free vase; (c) after pruning to be trained to vase.

Figure 5.9 Olive plant trained to bush (a) before and (b) after a light pruning in the 7th growing season after planting.

form a vase (see also Chapter 8) by selecting the primary branches and favouring their growth upward (Fig. 5.8c). In the following years the trunk will be cleaned from any lateral structure that may impede the attachment of the shaker's clamp, unless trees are going to be picked by hand.

A valid alternative to contain plant size and reduce pruning costs is to train plants as shown in Fig. 5.6a. Tipping of the arrow (cut no. 2) is done to promote growth of shoots in the lower part of the canopy, but it can be avoided in all those cases when good lateral shoots are already present. The plants so prepared will not be suitable for mechanical harvesting by shakers.

If the permanent skeleton does not have to be rigidly prepared, cuts must be reduced to a minimum in the first 4–7 years. Only those few shoots that can compromise the definitive shape and balance of the adult plant will be gradually eliminated. In particular, cuts must be made to eliminate: (i) overlapping distorted branches for good spacing between scaffolds and balanced plants; (ii) excessively low-hanging shoots to ease cultural practices on the row; (iii) shoots growing upright in the central or inner part in the canopy.

If shoots are well distributed in the canopy, pruning during training can be virtually neglected. The plant in Fig. 5.9 was trained to a completely free canopy and received the first pruning on the 7th year after planting. The cumulated yield over the first seven growing seasons after planting was 56 kg of fruit (equal to 16 t per ha) without any cost for pruning. Pruning on the 7th year removed only two major branches in the central part of the canopy and two shoots in the basal part. This plant can be gradually converted into a vasebush with little further expenditure since all the pruning can be done quickly from the ground every other year. To

Figure 5.10 Young plants during the training period. Plants are trained to form the following systems: (a) vasebush or bush (cv. Picholine); (b) central leader (cv. Kalamata); (c) single-trunk free-canopy (cv. Frantoio).

convert the bush into a globe, again little pruning is necessary. The globe does not need any shoot to be eliminated in the central part and the first pruning can be done when the plant is mature (8–10th year) to reduce height (sections 8.2; 8.6).

In summary, if free-canopy training systems (vasebush, free vase, STFC) are adopted the only difference in pruning lies in the presence of the single trunk free of lateral shoots or branches. The first pruning must be done not earlier than the fourth year after planting (unless plants grow unbalanced) and only consists in removing a few overcrossing shoots and watersprouts to allow more light penetration in the central part of the canopy. Pruning must be done before plants start showing signs of exhaustion in the lower part of the canopy. Figure 5.10 shows young plants trained to different systems (see Chapter 8 for details).

5.6 Different cultivars

The number of olive cultivars listed worldwide is huge (Bartolini et al., 1998). The olive orchard usually includes two or more cultivars to meet the requirement of cross pollination. The choice of the cultivar is one of the most delicate aspects for the success of

Figure 5.11 The effect of cultivar on growth habit of olive plants during the training phase. The plants are trained to free-canopy systems. Canopy shapes: (a) open (cv. Leccino); (b) pendulous (cv. Pendolino); (c) erect with clumped foliage on the distal part of shoots (cv. Moraiolo).

the orchard since it influences both quantitative and qualitative results. In general, cultivars native of the area where the orchard is located are to be preferred for optimal adaptation to climatic conditions and for production of typical fruits and oils.

Cultivars differ in vigour, growth habit and product use (see also sections 3.2 and 6.4). Pruning should be adjusted to the cultivar starting from the training phase. Figure 5.11 illustrates plants with an open, expanded, or pendulous growth habit. In pendulous cultivars pruning must favour the development of erect shoots by eliminating the more pendulous ones (Fig. 6.3). For single-trunk systems it is often best to tie the central stem to a trainer stick.

In cultivars with an erect habit, shoots and branches tend to have a narrow crotch angle. Thus, pruning must be aimed at opening the canopy by favouring the shoots oriented toward the external part (Fig. 6.3). The central part of the plant must also be pruned more heavily for light penetration. Individual secondary and tertiary branches should be left free to grow and produce so that cropping brings shoots down.

Cultivars with an open canopy are easier to be pruned. Since vegetation occupies most of the available volume it is not difficult to achieve the right inclination by cutting. Pruning cuts will be directed to widen or make more narrow the angle of inclination of branches with respect to the main stem.

Chapter 6

Pruning mature trees

6.1 Objectives

The objectives of pruning trees in full production are to: produce high yields of high quality, stimulate vegetative growth of fruiting shoots, maintain skeleton structure, prevent ageing of the canopy, eliminate the dead wood, improve air circulation and light penetration to make the canopy microclimate unfavourable for the establishment and thriving of pests and diseases. Primary and secondary branches must be shortened to maintain as rigid a skeleton as possible for mechanical harvesting. Machinery for cultural practices and harvesting should have enough room to be moved without difficulties.

Pruning should be also aimed at maintaining the canopy within a size compatible with economical management of the orchard. However, the expansion of the canopy in an adult plant can be restricted by pruning only to a certain extent, because size and vigour depend on genetic factors, climatic and cultural conditions. If the plant is pruned too severely it will respond with an excess of vegetative shoots and watersprouts, which will inhibit fruiting partially or completely.

6.2 Techniques

The concepts previously discussed for pruning young plants hold for mature trees as well (Fig. 4.2). The main operations routinely used for pruning mature trees are listed in Table 6.1. Primary branches are pruned individually, as if each branch were a single unit of production. Primary branches are replaced only if they are old or badly damaged (see sections 6.8 and 6.9).

Table 6.1 List of annual pruning operations on mature olive plants.

1   Thin shoots in the upper part of the canopy;
2   Identify the maximum height at which the tree will be allowed to grow and cut the main axis at the point of insertion of a secondary branch;
3   Eliminate suckers and watersprouts;
4   Shorten secondary branches to contain the lateral expansion of the canopy;
5   Eliminate exhausted shoots and renew secondary and tertiary branches;
6   Eliminate the vigorous shoots inserted with a narrow angle on the primary branches and those inserted at narrow spacing or overlapping;
7   Remove damaged shoots and branches.

The fruiting surface is renewed by either thinning individual shoots, or by the suppression of entire secondary or tertiary branches once most of their fruiting shoots have been exploited (see also Chapter 4). In modern pruning the selective thinning of individual shoots should be kept to a minimum because it is time-consuming. Thinning of shoots consists in eliminating a number of fruiting shoots from the branch every year (or every two years at a maximum) so as to distribute the remaining shoots evenly. In this way, the original inclination of the branch is maintained. Fruiting branches should be shortened every two or three years to prevent their excessive elongation. For practical reasons, when pruning is severe or done with pneumatic tools or a chain saw only few large cuts are made on secondary branches rather than suppressing individual shoots. Traditional pruning techniques require that secondary branches be maintained. In free-canopy systems, where there are no proper secondary or tertiary branches, shoots and branches are removed once they have been exploited.

Entire secondary and tertiary branches are suppressed if they are not adequately spaced, if they are overlapping, damaged, or most of their fruiting shoots are exhausted. The exhausted part must be cut off, leaving enough mixed shoots for the current year's production. Vegetative shoots, including watersprouts and suckers, must be removed periodically because

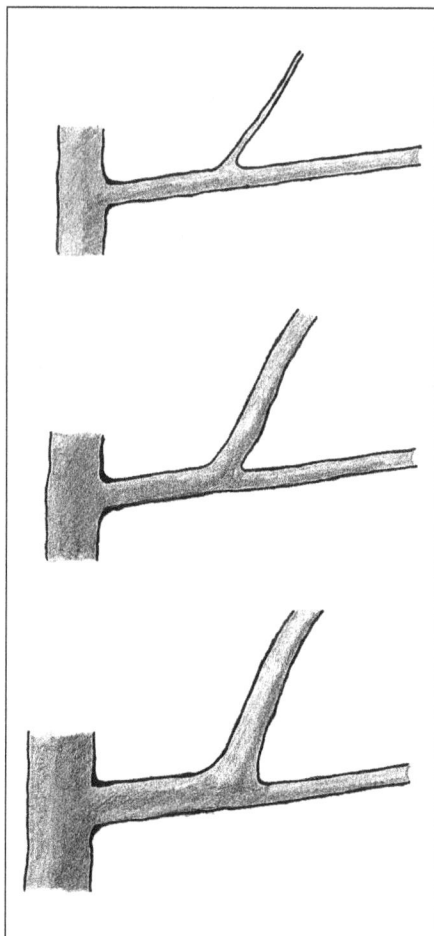

Figure 6.1 The differential growing of an erect shoot inserted on a horizontal shoot. The drawing shows the growth in three successive growing seasons (from top to bottom).

they always grow more vigorously than fruiting or mixed shoots (Figs 3.11; 3.12; 6.1). Watersprouts also contribute to make the inner part of the canopy dense and shaded, but can be useful to rejuvenate old trees (section 6.8).

Different examples of pruning a fruiting branch are shown in Fig. 6.2. Case (a) is an example of light pruning, whereby every productive unit (or lateral shoot) is pruned to eliminate the exhausted shoots as well as those inserted vertically on the upper side of the branch. The branch length is partially reduced by cutting the distal part of the shoots. Pruning type (a) can be done annually and assures a good control of plant shape and cropping, but it is time-consuming and costly.

In (b), the pruning is more severe than in (a), with fewer cuts on older wood. Pruning is performed with few large cuts, leaving vegetative shoots on the upper

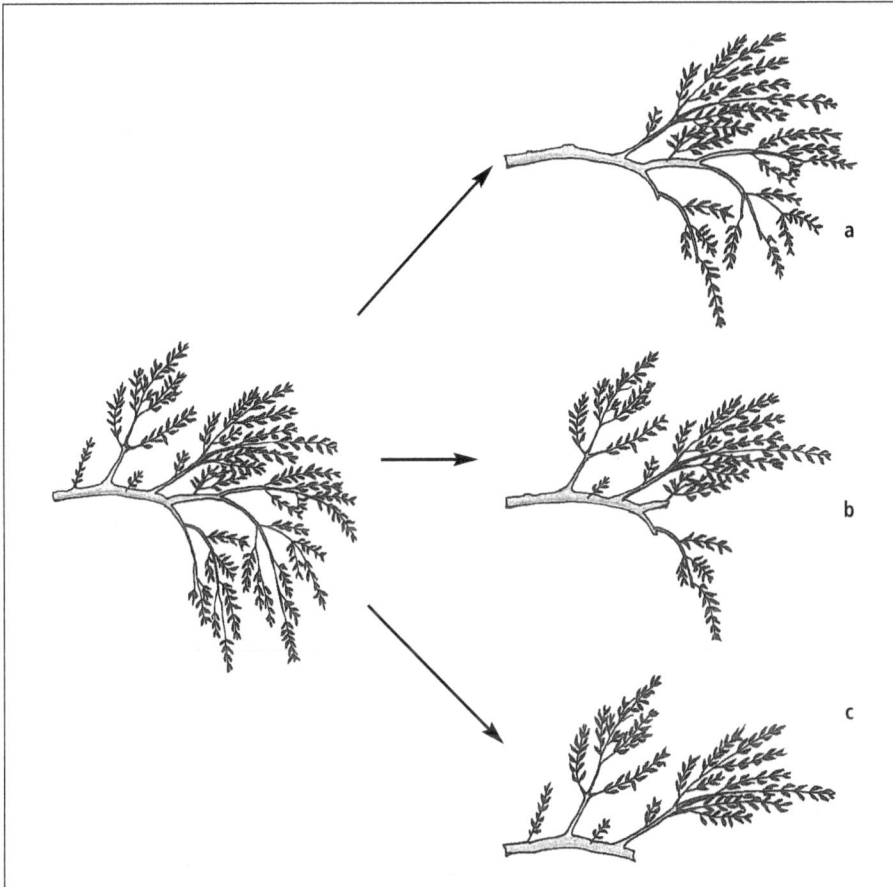

Figure 6.2 Schematic representation of pruning a fruiting branch at different levels of severity: (a) only few shoots are removed; (b) most pendulous shoots are removed and branch length is reduced; (c) all pendulous shoots and the terminal part of the branch are removed.

side, whereas most pendulous shoots are removed. The branch is shortened to keep fruiting closer to the centre of the plant and stimulate vegetative growth. The large number of mixed shoots on the upper side of the branch assures the renewal of vegetative growth. This method is commonly used when trees are pruned every other year.

The heaviest pruning is illustrated in (c). All the pendulous shoots and the terminal part of the branch are eliminated by pruning. The fruiting surface of the branch is renewed with fruiting concentrated in the mixed shoots on the upper side of the branch. The shortening of the branch stimulates emergence of new vegetative shoots and improves light penetration. Since only one large cut is needed, this method is also suitable for pruning at low frequency with a chain saw or pneumatic shears. It is not necessary to eliminate the sprouts from the upper side of the branch because they will be useful for renewing vegetative growth (Fig. 6.2).

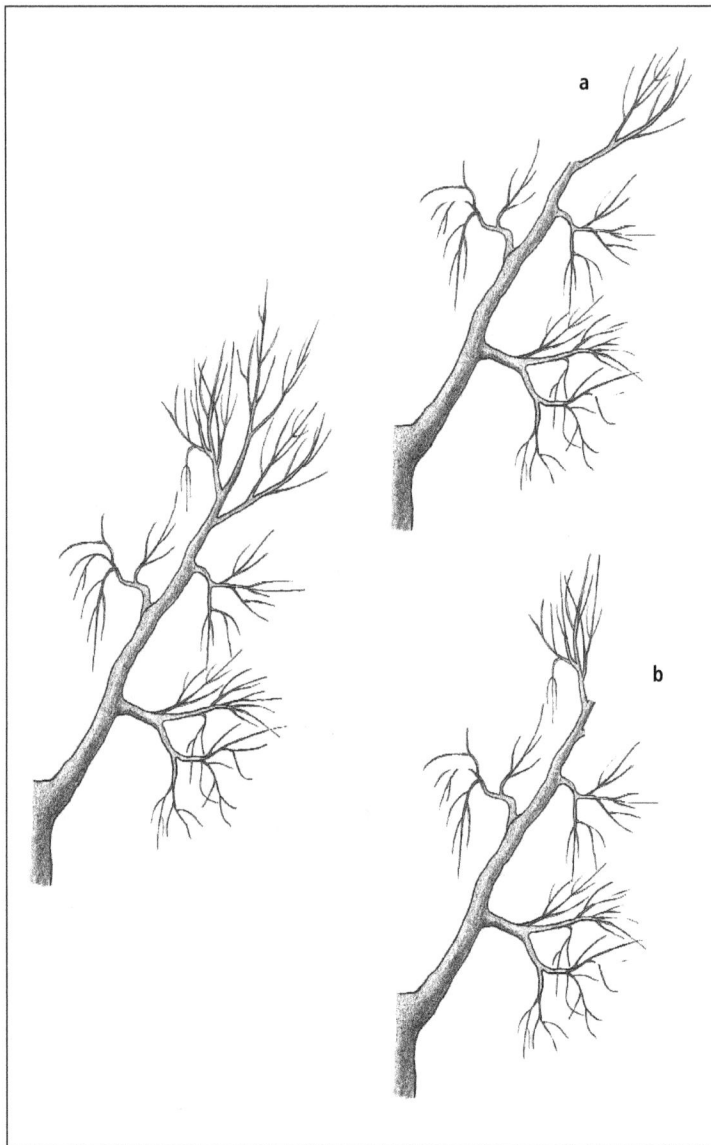

Figure 6.3 Cut of the terminal part of the primary branch (left) of a vase system. The cut is directed to counterbalance the natural growth habit. A lateral shoot is left externally (a) or internally (b) to replace the role of the suppressed axis.

In all three examples in Fig. 6.2, the goal of pruning is to eliminate exhausted parts, maintain some fruiting shoots, and stimulate emergence of vegetative shoots. However, the result of the three pruning treatments is different: fruit yield and branch length are decreased progressively from (a) to (c), whereas vegetative activity and light penetration are increased. These three pruning methods can alternatively be applied to different branches or over the entire canopy depending on the training system, frequency of pruning, and cultural conditions.

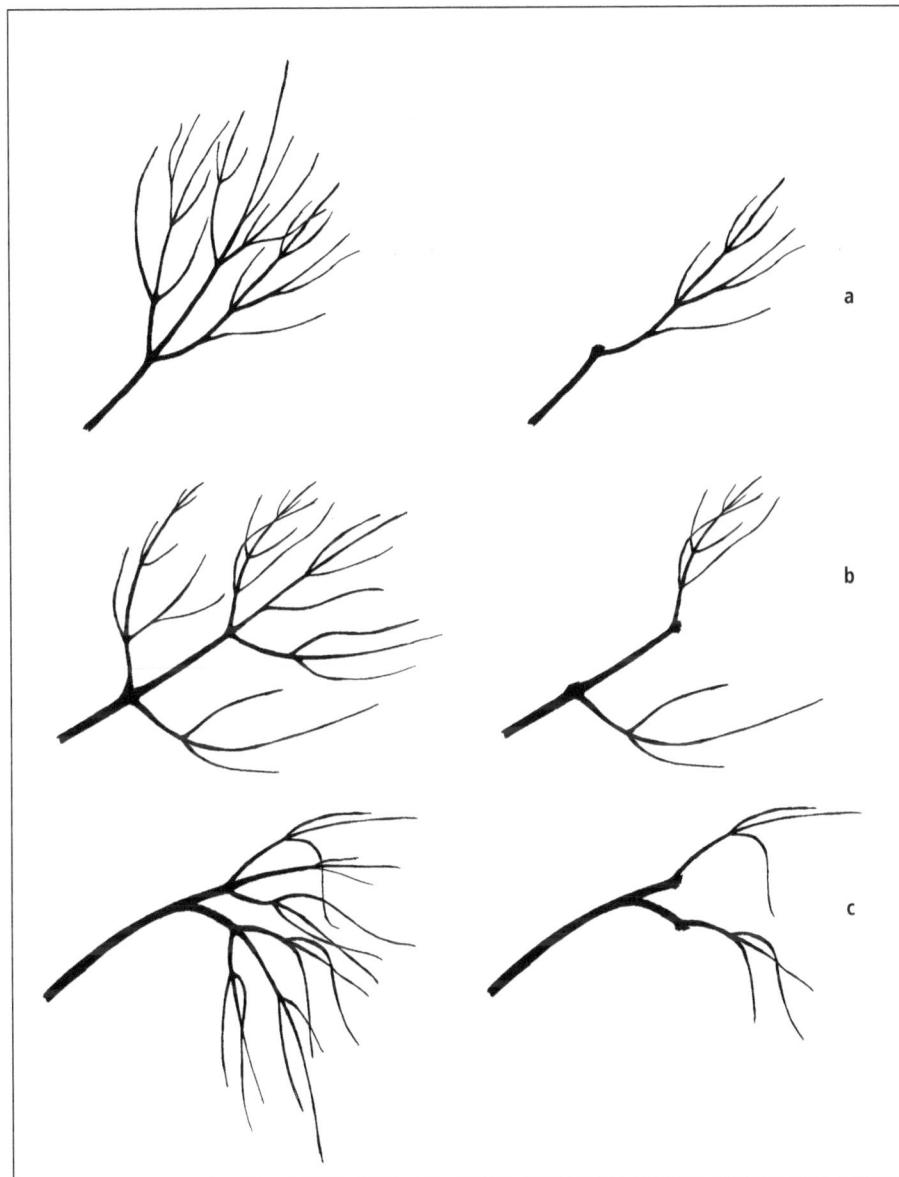

Figure 6.4 Schematic drawing showing examples of pruning cuts on shoots with different growth habits: (a) erect; (b) expanded; (c) pendulous.

Long branches with a wide angle of inclination do not transmit well the vibration by the shaker and hinder the movement of machinery along the row. In general, the length of secondary branches should be inversely proportional to the number of primary branches present on the plant. If primary branches are numerous (five or more), the secondary branches must be kept relatively short by heading cuts. Heading back of primary branches must be done at the point where a healthy and well-sized secondary branch originates (the diameter of the secondary branch should not be less than half the diameter of the cut branch). In this way, the number and vigour of the new shoots arising from near the cut surface are controlled by the shoots remaining on the secondary branch.

The heading back of primary branches can also be used to adjust their inclination. Pruning should leave the shoots oriented towards the outside or the inside of the canopy, depending on whether a wider or narrower angle is desired (Fig. 6.3). Selective cuts next to a lateral on the upper or lower side of a primary branch allow the pruner to regulate the inclination of the branch itself, depending on the growth habit of the cultivar (Figs 6.3; 6.4; sections 3.3 and 5.6). Canopies with an expanded habit are more easily manageable by pruning. In pendulous cultivars vertical shoots oriented towards the centre of the canopy will preferably be left, whereas in erect cultivars shoots oriented towards the outside must be favoured (Fig. 6.4). Since in cultivars with an upright habit the centre of the canopy becomes dense, pruning will have to suppress some of the branches for better light penetration and form secondary branches with a wide angle. However, the excessive inclination of branches produces a marked emergence of new shoots in the inner part of the canopy and shading underneath.

6.3 Crop load

The effect of pruning on alternate bearing is generally small. Morettini (1972) reported that pruning had no effect on alternate bearing of 'Canino' trees over 13 production years. More recently, Morettini's findings were confirmed in a trial where 'Frantoio' plants were grown unpruned until the 6th year after planting, and then two pruning regimes were established: no pruning or annual pruning (vasebush). Annual pruning determined a lower cumulated yield over five years of production (67% of yield of the bush), but no effect on the alternate bearing habit (Fig. 6.5). Alternate bearing is not prevented even when trees have been pruned accurately every year since planting. In an irrigated orchard where trees were planted at 6 x 3 m and trained to a central leader, alternate bearing was similar to trees that had been grown unpruned for six years

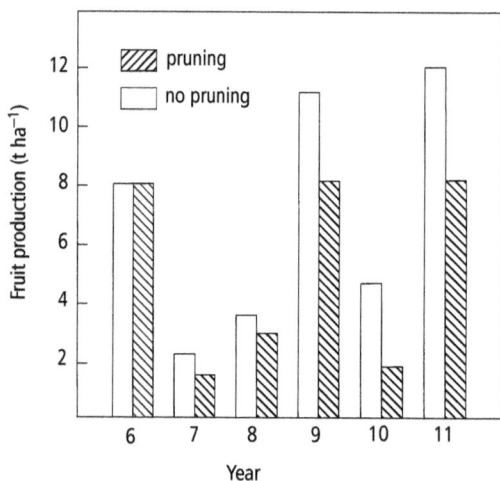

Figure 6.5 The effect of annual pruning on yield of trees (cv. Frantoio) planted at 6 x 6 m in 1986. All plants (about 130) were left unpruned until spring 1991 when one to three branches were removed from the central part of the canopy. Then, half of the plants were left unpruned (no pruning) and the other half were pruned annually to vasebush.

and then trained to a bush. Annual yields of plants pruned annually, biennially or every four years in Spain also showed that the different frequency had negligible effects on the pattern of alternate bearing (Fig. 6.6; Garcia-Ortiz, 1998; Pastor, 1989).

Fruit thinning has proven a valid method to control the size of the crop and reduce alternate bearing in *Citrus* (Monselise and Goldschmidt, 1982) and it has been recommended for olive as well (Martin et al., 1994a), but this expensive practice is feasible only for table olives. Pruning should be light in the spring following the "on" year, and more

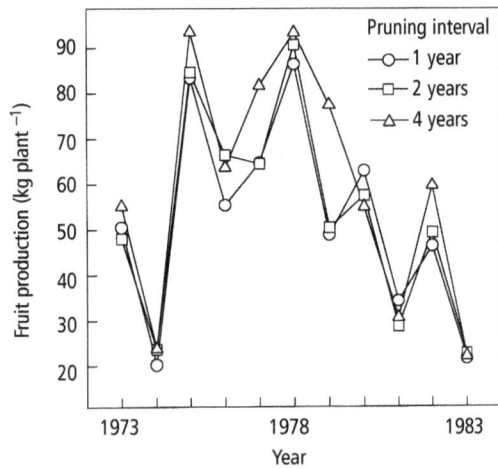

Figure 6.6 The effect of a different frequency of pruning on yield of mature olive trees in Spain. Note the lack of an effect of pruning on the sequence of yield fluctuations (drawn from data reported in Pastor, 1989).

severe after the "off" year. If pruning is done after bloom, it will be more effective to decrease the number of flower buds on the tree by selectively removing the shoots bearing most fruits.

Although adjusting the intensity of pruning to crop load is not difficult to implement, it is seldom done in practice. Pruners tend to do the same pruning every year regardless of the previous year's yield. Moreover, it is not uncommon to have high- and low-cropping trees in the same orchard, which complicates the work if selective pruning has to be performed. Finally, reducing markedly the crop load of trees with a potentially large crop is a hard decision to make.

A reduction in alternate bearing per hectare can be obtained by dividing the orchard in small units that are pruned at intervals longer than yearly. In practice, the number of plants must be divided for the duration of the pruning cycle (two or three years): the first third of trees (or half) will be pruned in year one, the second in year two, and so on. The amplitude of yield fluctuations is decreased because the lots pruned in different years are likely to produce asynchronously.

6.4 Product destination

Olives are produced for either table or oil production. There are only few cultivars that can be considered dual purpose, but because qualitative features for oil or table production diverge somewhat, it is difficult to achieve maximum quality for both table and oil production with just one cultivar. The management of the olive orchard, including pruning and the choice of the training system, varies according to product use.

Fruits for oil production must have a high flesh-to-pit ratio, a high oleic-acid/other-fatty-acids ratio, and a high content of polyphenols and antioxidants. The presence of chlorophyll is important because it confers the oil a green colour, which is appreciated by consumers. A large fruit size improves yield to mechanical harvesting. Pruning mature trees of cultivars for oil production is relatively easy. The effect of pruning on oil characteristics is minor, since oil quality depends mainly on the cultivar, the stage of harvesting, the time between harvest and processing, and processing technology. Oils of excellent quality and taste can be obtained with many different pruning methods as long as harvesting and processing are performed correctly. Even trees unpruned for 10 years produce top quality oil (Cantini et al., 1998). Morettini (1972) reported that the cumulated production of individual plants of 'Frantoio' and 'Moraiolo' was about 150% higher in unpruned plants than in pruned trees over 15 years of production.

For table olives the fruit has to be large with a high flesh-to-pit ratio, high soluble carbohydrates content, and low oil yield. The colour of table olive is determined by the cultivar, the stage at which fruits are harvested, and the curing process. More details on production and processing of table olives are found in Garrido Fernandez et al. (1997) and Brighigna (1998).

There is an inverse relationship between the number of fruits per tree and the mean fruit weight (Garcia-Ortiz et al., 1997). If yield is reduced, fruit quality is increased, which results in higher revenues. The maturation process is also more uniform if the crop load is not excessive. Thus, pruning is used to reduce crop load and improve light penetration. The number of shoots and fruits that need to be removed should be adjusted so as to obtain a good commercial crop.

Cultivars for table production should be pruned annually. The pruning must be accurate and more severe than for oil cultivars. Summer pruning should be performed between 20 and 30 days after full bloom to be effective on fruit size (Rallo, 1997). The technique for renewing fruiting shoots that is adopted more often is to thin many of the individual shoots rather than making cuts on few large branches. Shoots growing in the inner part of the canopy, and especially those on the upper side of primary branches, should be eliminated to improve light penetration and air circulation. Olive fruits in well-pruned and illuminated parts of the canopy grow bigger and develop a more intense colour with a higher sugar content than fruits in shaded or unpruned parts. It is preferable to have training systems where light and fruits are distributed evenly within the canopy and the foliage is not dense (e.g. vase, vasebush, central leader).

6.5 Different training systems

The pruning of mature trees varies according to the training system since one of the main objectives is to maintain the tree structure over the years. The different training systems are described in Chapter 8. Here only the main guidelines for pruning different systems are reported.

Canopy architecture

The olive canopy is formed by a woody skeleton bearing the one- and two-year-old fruiting shoots. The skeleton represents the permanent structure and consists of different parts (Fig. 6.7). The trunk (T) is the main axis of the plant. The primary branches (B1) originate directly from either the ground or the main trunk. Secondary branches (B2) are inserted onto primary branches; tertiary ones (B3) originate from secondary branches, and so on. Orientation and length of primary and secondary branches vary according to the type of pruning and training system. Once the structure of the plant is fully formed, pruning is used to renew the fruiting surface and contain the size of the mature tree.

The following zones can be identified on a fruiting branch, depending on the prevalence of different types of shoots (Fig. 6.8; see the box on pages 29–31 of Chapter 3 for classification of shoots):

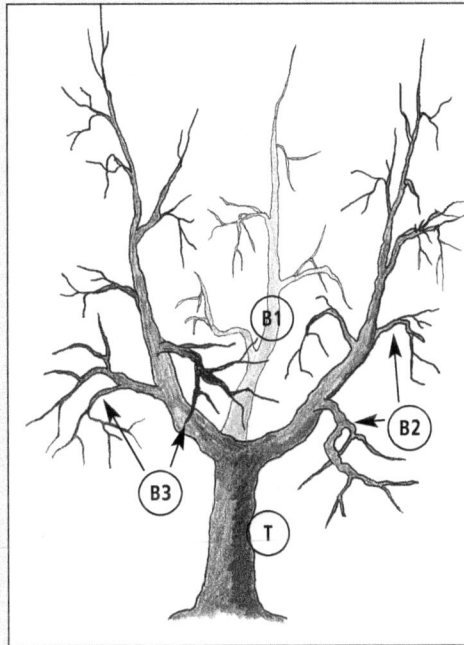

Figure 6.7 Schematic representation of the woody skeleton of an olive tree trained to vase: trunk (T); primary branch (B1); secondary branch (B2); tertiary branch (B3).

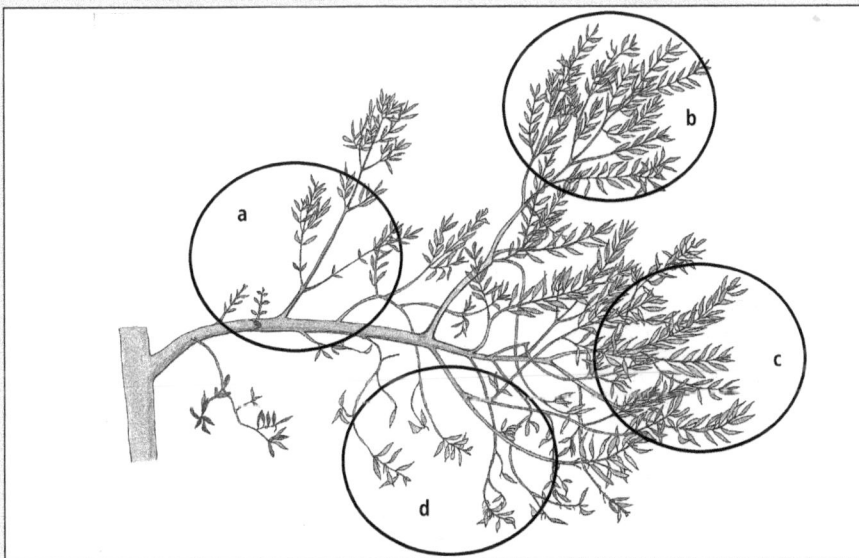

Figure 6.8 Schematic drawing of a fruiting branch of a mature olive plant showing four zones differing in the prevailing type of shoots. (a) vegetative shoots; (b) mixed shoots; (c) fruiting shoots; (d) exhausted shoots.

(a) zone with vegetative shoots. Vegetative shoots are usually found in the upper, proximal part of the branch. Shoots originating in this zone are vegetative for one or two years before they start bearing flowers; they grow almost vertically and often develop into vigorous watersprouts. Vertical shoots tend to overgrow all those shoots present in the area distal to their point of insertion (Fig. 3.11b). Erect shoots are useful to rejuvenate the branch once it is exploited.

(b) zone with mixed shoots. Mixed shoots are usually found in the upper part of secondary or tertiary branches and their contribution to yield is normally substantial. One-year-old mixed shoots grow vigorously (about 0.3–0.4 m) and this often causes the deprivation of shoots and fruits in the proximal part of the branch. Mixed shoots should be thinned only if the foliage becomes dense; it is important not to eliminate too many of them as they will be useful to renew the fruiting potential of the branch in the years that follow. Small branches will be shortened to maintain fruiting close to the centre of the canopy. A few of these branches will be suppressed if they are too numerous or closely spaced.

(c) zone with fruiting shoots. Fruiting shoots are present in the outermost layer of the canopy. Pruning can be done essentially in two ways: (i) thinning fruiting shoots to reduce the competition among them, but leaving the entire length of the branch, or (ii) heading back the main branch. The former method must be followed in cases when the original position of the branch is to be maintained, in the latter case the branch is shortened to control the lateral expansion of the canopy.

(d) zone with exhausted shoots. Fruiting shoots become exhausted after they have produced abundantly for at least two years. Exhausted shoots are easily identified by their weak growth and the few leaves present at the end of long, bare stems (Fig. 6.9; see also section 3.4). Their section is smaller than shoots with more active terminal growth. Exhausted shoots are found more frequently on the lower side of secondary and tertiary branches in the lowermost part of the canopy where little light penetrates.

Figure 6.9 Exhausted fruiting shoots growing in the shaded part of the canopy. The shoot was photographed in the same position it had occupied originally in the canopy.

In systems suitable for hand-picking, management costs of the tree are reduced by preventing plants to grow high; to achieve this, primary branches are periodically shortened and the shoots in the apical part are thinned more severely than those growing in the lower part of the canopy. In plants trained to vase, the shoots originating from the primary branches and directed towards the centre of the canopy are preferably eliminated since they tend to block light penetration through the canopy. Primary branches are headed back to a lateral shoot or branch to decrease tree height. Less emphasis is given to the structure formed by secondary and tertiary branches in current types of vase systems than in the past. Hence, renewal of the fruiting surface will be done mainly by elimination of entire fruiting branches once they have been exploited, rather than elimination of individual fruiting shoots.

Figure 6.10 Thinning of the lateral shoots to favour the elongation of the arrow in a tree trained to central leader at a 6 x 3 planting distance (courtesy of G. Paiardini).

The thinning of shoots in the central part of the canopy of trees trained to globe will be lighter than in open vase systems. The internal part of trees trained to globe or free-canopy systems often becomes very dense. Shoots growing in this zone, and especially those on the upper side of primary branches, should be thinned to favour light penetration and improve air circulation. Only in extreme cases of canopy density is it necessary to thin the secondary branches to open gaps and give more space to the remaining branches.

In central leader systems, heading cuts should be made at the terminal end of the main axis to reduce the height of the central leader. An upright shoot will be selected as the new arrow of the tree. Heading-back cuts are also performed to shorten lateral branches and to keep a relatively rigid structure for mechanical harvesting as well as to keep foliage and fruiting close to the main axis. The vertical shoots inserted on the primary branches growing close to the trunk should be eliminated to avoid crowding and maintain a wide angle of inclination of the branch (Fig. 6.10). Suckers, shoots inserted low on the trunk, and the low-hanging shoots should be eliminated in single-trunk systems because they will slow down mechanical harvesting.

Figure 6.11 Heading of central leader trees to form a single-trunk vase with four primary branches (courtesy of R. Guerriero).

The opening of gaps is essential for table cultivars, whereas a certain degree of shading is tolerated in oil cultivars. However, vigorous shoots in the central part of the canopy should be left in climates where the high levels of sunlight may damage the bark of scaffold branches. Full canopy systems (e.g. globe, bush) are recommended in these conditions.

There can be specific cases when one is unsatisfied with a particular training system, but while the orchard is still young it is not worth eradicating it. Trees can be pruned to modify the woody frame. An example is illustrated in Fig. 6.11, where central leader trees were cut, leaving only the lower part of the trunk, and three vigorous shoots were selected to form the primary branches of a vase.

6.6 Mechanical pruning

The initial tests on pruning olive trees by machine date back to over 30 years ago (Morettini, 1972). Nowadays, mechanical pruning is considered the obvious solution to reduce pruning costs and avoid the increasing problem of lack of skilled labour in many areas. Mechanical pruning is usually achieved with a sickle bar or circular saws mounted on a support at the end of a rigid arm carried by the tractor. Although several prototypes of sickle bars, differing in the number of blades, their arrangement on the bar, and the size of the cutting edge, have been proposed and tested so far, the principle of operation is similar. The blades cut the external edge of the olive canopy while the tractor moves at low speed (less than 2 km per hour) along the row. The blades can cut horizontally on the top of the canopy (topping), vertically or at an inclination (hedging). The resulting canopy profiles assume the shape of cones, pyramids, or parallelepipeds. Pruning performed with the bar inclined to form a narrow angle with the vertical axis is the most commonly used method in both traditional and modern systems.

The effect of mechanical pruning on yield has been studied in many countries. The most extensive project has been carried out in Spain where several combinations of mechanically-pruned treatments were tested on single-trunk trees (cv. Picual) planted at an 8 x 4 m distance in 1963 (Pastor, 1989). Mechanical pruning started in 1981 and included five different treatments whereby hedging and/or topping cuts were performed every two, three or four years. The control trees were pruned manually approximately every other year (seven times in 12 years). The results showed that plants pruned mechanically yielded at least 91% of those pruned manually (cumulated over the period 1981–87). Interestingly, the mechanically pruned trees that yielded the most (100% of the control) were those receiving the least pruning. In the latter treatment, hedging and topping cuts were performed either every five years or topping was followed by a year of no pruning and then one of hedging only (Pastor, 1989).

In another study mechanical pruning was applied to 25-year-old trees trained to three trunks and irrigated during the summer (Pastor et al., 1991, cited in Garcia-Ortiz, 1997). The accumulated yields of mechanically pruned trees were not different from those pruned manually, and trees that received a light pruning by machine yielded better than more heavily pruned ones.

In Italy a few studies have been conducted on both intensive orchards and old trees in the last 15 years. Giametta and Zimbalatti (1997) found negligible differences in yields of 10-year-old trees (cv. Leccino) in the three harvests following mechanical pruning. Mechanical pruning took only 4 man h/100 trees, about 3% of that required for hand-pruning. It should be noted, though, that the reported 128 man hours required to prune 100 trees manually is excessive, and indicates that hand-pruning was not done according to modern principles of pruning. Pochi et al. (1996) studied the effect of tractor speed and blade rotation speed on containment of tree size and quality of cuts in a high-density orchard (5 x 2.5 m), but did not report yield data. Fontanazza et al. (1998) reported that mechanically-pruned trees (cvs. Leccino and Frantoio), planted at a 6 x 3 distance and trained to central leader, grew and produced more than manually pruned trees. Mechanically pruned trees yielded 8.9 kg, whereas hand-pruned trees averaged 6.2 kg fruit per tree over a 10-year period.

Undoubtedly, mechanical pruning allows the pruner to save considerably in annual cultivation costs. Fontanazza (1993) reported that over 80% of pruning costs could be saved by mechanical pruning, which took only 2 hours of work by machine per hectare. More recently Fontanazza et al. (1998) reported that only one machine-hour per hectare was needed (excluding the time needed for turning the tractor at the end of each row). Giametta and Zimbalatti (1994) showed the advantages in terms of pruning costs and labour productivity by hedging and topping century-old trees of 'Ottobratica' and 'Sinopolese'. In a different study Giametta and Zimbalatti (1996) also described machinery and methods to prune very large trees (even taller than 20 m).

Mechanical pruning is still uncommon in olive growing worldwide. This is mainly due to some practical problems. First, unlike manual pruning, mechanical pruning is not selective as it tends to uniformly cut the surface rather than open gaps in the canopy. The indiscriminate cutting causes the heading of shoots on the most external and well-illuminated part of the canopy, which results in a decrease of yield and fruit size, and the formation of too many short vegetative shoots suckering from the adventitious and axillary buds next to the cut surface. Vitagliano (1969) reported that hedged sides of trees (planted at 5 x 2 m) had shorter internodes, fewer flowers per inflorescence, more aborted ovaries, and less fruit production than unhedged sides. Another problem of mechanical pruning is that it eliminates neither the shoots inserted on the older branches, nor the vertical shoots in the shaded and inner part of the canopy.

If olive trees are repeatedly pruned mechanically they will assume the appearance of thick regular hedges of scarce productivity. A mat of shoots forms on the outer edge, which prevents light penetration in the inner and lower parts of the canopy. Moreover, periodic pruning will be necessary to restructure the canopy, which implies drastic cuts and one or two unproductive years. Mechanical pruning is less suitable on old trees (weak stimulation of new growth), and for the production of high-quality table olives. An additional problem to the expansion of mechanical pruning lies in the characteristics of orchards (small size, steep slopes) in many olive-growing areas.

Because of the above reasons mechanical pruning is not sustainable in the long run because it determines an excessive ageing of the canopy. Experimental evidence shows that mechanical pruning by itself is not very successful, and that it has to be alternated with hand-pruning. Manual pruning integrates mechanical pruning by selectively eliminating the dead wood and opening gaps in the canopy.

There is no fixed time interval for the optimal strategy. In general, the period of rotation between manual and mechanical pruning ranges from three to six growing seasons. Garcia-Ortiz et al. (1997) recommended an interval of four years between two successive mechanical prunings. The combination of mechanical pruning the first year, followed by one or two years of manual pruning or no pruning and then manual pruning again has proven successful in Spain (Pastor, 1989). Fontanazza et al. (1998) suggested pruning mechanically the first year, not pruning the second year, and pruning manually the third year trees (cvs. Frantoio and Leccino) planted at 6 x 3 m and trained to central leader. They also suggested reforming the canopy structure drastically every nine years. Pastor (1989) recommended pruning mechanically the first time after a high cropping year and eliminating at least 0.75–1.0 m of the outermost shoots.

Besides the specific protocols that have been tested so far, it is clear that strategies to integrate manual pruning with mechanical pruning have to be developed at the farm level. To determine the optimal rotation of mechanical and manual pruning, the productivity and vegetative growth of the orchard have to be considered. Hedging can be done only on one side of trees in a row in year one; the other side will be hedged the next year. To reduce costs further, one or two years without pruning should be included in the rotation. Finally, it is safer not to prune mechanically before the trees are at least 8 to 10 years old. Younger plants may respond too vigorously to the unselective heading of shoots, making the final canopy structure more difficult to achieve.

6.7 Adapting traditional systems to mechanical harvesting

Traditional training systems are still common in olive growing worldwide. In many instances trees are very productive and it is not worthwhile eradicating the grove and planting a new orchard. Because of the size and age of trees, hand-picking can be extremely expensive and has to be replaced by mechanical harvesting. However, the yield efficiency to harvesting by machine is low. Pruning would then be specifically aimed at adapting the canopies to mechanical harvesting.

In a trial conducted on trees (cv. Frantoio) trained to polyconic vase at a 9 x 9 m planting distance, Vitagliano et al. (1983) showed that the partial or total elimination of pendulous vegetation improved yield by mechanical harvesting. A light or a severe pruning, whereby 40% or 80% of pendulous shoots were eliminated and secondary and tertiary branches were shortened, resulted in an increase of yield efficiency to harvesting by shakers with respect to the control (Table 6.2). The control trees were pruned annually by heading cuts to shorten the primary branches and maintain pendulous shoots for easier hand-picking. Labour productivity was increased from 317 (control) to 363 and 382 kg of fruit harvested per hour per worker for the severe and light pruning treatments, respectively. Selective pruning also increased the amount of wood removed, reduced the time needed to prune a single tree, and decreased canopy volume (Table 6.2).

Table 6.2 The effect of pruning to adapt old trees to mechanical harvesting by shakers. Light pruning consisted in the elimination of about 40% of pendulous shoots; severe pruning consisted in the elimination of about 80% of pendulous shoots and shortening of most secondary and tertiary branches. The control was pruned as a traditional polyconic vase (modified from Vitagliano et al., 1983).

	Light	Severe	Control
Time for pruning (min/plant)	20	28	33
Wood removed (kg/plant)	34	47	22
Canopy volume (% of control)	87	65	100
^Yield (kg/plant)	34	43	37
Yield efficiency (% of total)	78	76	68
Time for harvesting (sec/plant)	296	372	231
*Labour productivity (kg/h/person)	382	363	317

^ Mean of yields over five years.
* Includes time for laying the nets and recovering the product.

Improvements in yield and yield efficiency to mechanical harvesting were also obtained by Tombesi and Jacoboni (1974) and by Santos (1988; cited in Pastor, 1989). In the latter work, the percentage of fruits detached by shakers was 86% in trees with branches more erect and rigid where most pendulous shoots had been eliminated by pruning *versus* 79% of trees pruned traditionally.

In California, hedging trees mechanically has been used to prepare the canopy of trees (cv. Manzanillo) for harvesting by a comb-whipping machine (Ferguson et al., 1999). The hedging treatment did not decrease the percentage of fruits removed mechanically, whereas the topping treatment reduced yield and value of fruits. Cull rates of black-ripe table olives were only 1% higher in hedged trees than in trees harvested manually (Ferguson et al., 1999).

To adapt trees to mechanical harvesters, pruning consists essentially of removing the branches and shoots not directly inserted on the primary branches. Tertiary branches are eliminated and secondary branches are renewed more frequently than with traditional pruning methods. Moreover, not all secondary branches are maintained. Those inserted at a wide angle (an angle of 45° is considered optimal) are preferably eliminated as they do not transmit the vibration efficiently. Most pendulous shoots are removed. Lateral shoots should be eliminated from the trunk for at least 1 m from the ground. Since pruning is time-consuming, the economics of adapting the canopy to mechanical harvesting should be carefully evaluated.

Not all old trees can be adapted to mechanical harvesting. Suitable trees are those presenting a single trunk without lateral shoots in the first 1 m from the ground. Trunks should be healthy without signs of rot or they will not withstand the vibrations when shaken. Systems like the vasebush or the bush are unsuitable unless one wants to invest a lot of time and effort in pruning. Alternatively, the single primary branches of trees without a trunk can be shaken, but harvesting will be time consuming. In the latter case it is useful to reduce the number of primary branches to no more than four. Trees should also be spaced at wide distances to allow easy movement of machinery. Finally, plants of pendulous cultivars are more difficult to reform than those with an erect habit.

6.8 Old trees

Because of the longevity of olive trees it is not uncommon that the pruner has to face the problem of pruning old trees (50 or more years old). Old trees are often large, high, with one or more trunks unsuitable for mechanical harvesting and with an excess of old wood versus young wood (Figs 1.2; 6.12; 6.15). The pruning of old trees can be extremely expensive and should be done only if trees are so productive as to justify their rejuvenation. At any rate, old trees can remain productive and healthy for a long time if appropriately pruned and cultivated.

Pruning old trees mainly involves stimulating vegetative activity, decreasing the amount of old wood on the plant, and containing tree size. Thus, pruning for old

Figure 6.12 Severe pruning of a traditional vase. Note the cuts at the top of the canopy and the amount of wood removed on the ground.

Figure 6.13 Canopy of a mature olive tree (cv. Frantoio) rejuvenated from polyconic vase in two successive growing seasons. The tree has a high leaf-to-wood ratio and it is very productive. Since the rejuvenation pruning, the tree has been pruned manually with a chain saw every other year.

trees should be more severe than for young or mature plants. Although it is difficult to generalise, the amount of wood removed should approximately double what is usually taken off when pruning mature trees every year. As a rule of thumb, about 50% of the wood should be eliminated from old trees. The old wood should preferably be removed, leaving as much of the young wood as possible. The primary branches are shortened to promote the emergence of new vigorous shoots while simultaneously eliminating substantial amounts of old wood (Fig. 6.12). The secondary branches should be thinned if they are too numerous or overlapping. Fruiting shoots should be appropriately thinned to stimulate growth and production of those left on the tree. The remaining secondary branches should be shortened too; this will stimulate the burst of adventitious buds on the old wood. If the pruning is sufficiently severe the growth of new shoots will be adequate to make the old canopy productive for a number of years (Figs 3.1; 6.13). In this way, old trees can be pruned at a reasonable cost.

Figure 6.14 Rejuvenation of an old olive tree, whereby pruning has been performed in two steps. The young shoots to the left of the trunk (small arrows) are one year old, whereas the branch to the right (large arrow) is three years old.

However, in many instances pruning old trees involves the rejuvenation of the whole plant by leaving only the trunk and part of the old branches, while the rest of the canopy has to be reconstituted. In extreme cases, the entire canopy will be removed and suckers growing from the crown will be trained to form a new plant. This rejuvenation pruning is essentially done in two situations: (i) when trees are so large that their height needs to be considerably reduced, or (ii) when trees are badly damaged (see section 6.9).

There are several methods whereby trees can be rejuvenated. The primary branches should be shortened substantially to contain the final size of the tree to no more than 4.5 m. The primary branches are usually shortened at the insertion of a secondary branch. The reduction of canopy height and volume should be done as gradually as possible (Fig. 6.14), unless trees are in poor condition due to long-term neglect.

Since most old trees are trained as vase, rejuvenation usually consists in shortening only some of the primary

Figure 6.15 Mature olive tree trained to polyconic vase before rejuvenation of the canopy.

branches to the desired height, while leaving the other branches unpruned until new shoots form a sufficient canopy. In this way, the effect of the drastic alteration of the root/shoot ratio is attenuated. Then, the remaining branches should be shortened as well (Fig. 6.14) and the tree should be allowed to grow with minimal pruning in the year following rejuvenation. In vase systems with a long trunk, the watersprouts that will form the new branches are selected at a lower height on the trunk than the old primary branches. The selection of new branches will be appropriately spaced along the trunk, as explained in Chapter 5; they will be productive within four years from the rejuvenation cut. Subsequent pruning will remove only the damaged parts, suckers, and the most vigorous watersprouts, in particular those growing at the top of the primary branches and oriented towards the inside of the canopy. An example of an old tree trained to a vase system and rejuvenated by shortening the primary branches is shown in Fig. 6.16.

Pastor (1989) and Garcia-Ortiz et al. (1997) described the method adopted to rejuvenate old trees in the Jaen region of Spain. The method consists in the gradual rejuvenation of the canopy by removing one branch first, and then eliminating the rest of the canopy in the following years. This pruning is used as a management strategy to renew the canopy of old trees periodically and reduce to a minimum the pruning between two successive rejuvenation cuts (see also section 4.6).

In cases when the trunk and major scaffolds have been badly damaged by frosts, wood rot or other events, rejuvenation is achieved by cutting down the tree at the base of the trunk. In this way, a new canopy will be formed by the suckers emerging from the crown (see also sections 6.9 and 8.7). The removal of the entire trunk is not

Figure 6.16 Rejuvenation of an old olive tree. The canopy has been almost completely reformed. The canopy volume has been successfully reduced and the ratio between foliage and old wood is now adequate for high yields. Note the large size of new cuts made during the production pruning.

Figure 6.17 Rejuvenation of an old tree after frost damage. The old wood that had not been injured was left. The central part of the trunk shows signs of wood rot (courtesy of Dipartimento di Coltivazione e Difesa delle Specie Legnose, University of Pisa).

possible if trees are to be harvested by machine, plants have been grafted high, or the trunk has to be preserved for ornamental and landscaping purposes as in gardens or trees of historical interest (Figs 1.2; 3.1; 6.17). In these cases plants should be pruned so as to leave as much as possible of the old trunk and branches (sections 6.9 and 7.5).

Rejuvenation of the plant by cutting the trunk at the base assures excellent growth and productivity at minimum cost if trees are harvested by hand. In some traditional systems one or two suckers are sometimes left to replace the aged trunk, which is cut off only when the new shoots reach a certain size and start bearing fruit. This rejuvenation method is recommended only when a new tree cannot be planted or for a limited number of plants. In large olive orchards it is more convenient to cut the old trunks at ground level or uproot the old trees and plant a new grove.

6.9 Damaged trees

Frosts, pests and diseases are the most common causes of damage. However, olive plants can also be damaged by fire, hail, high light intensity, mechanical shakers, and erroneous cultural practices (e.g. inadequate application of herbicides, pesticides or fertilisers).

It is important to assess the extent of damage before pruning. An easy method is based on the age of the damaged organ. The damage can be considered: (a) light, if only leaves, one-, and two-year-old shoots are affected; (b) medium, if also the three- to five-year-old wood is injured; (c) severe, if primary branches and the trunk show symptoms. The root system usually remains viable, and the olive plant is almost always capable of resuckering from the crown. Only in extreme cases is the viability of the root system so impaired that the only solution is to plant a new tree.

Low temperature is the most frequent event damaging the above-ground parts of the tree. Frosts can be so common in relatively cold areas to be the limiting factor for the economic life of the orchard. The best time to assess the extent of damage is during the growing season following the occurrence of the event. Young trees are more susceptible to frost than old trees. In case of little damage (discoloured leaves often with a folded blade, but persisting on the plant) pruning will be light. If the cambium zone appears dry and brownish, and one- and two-year-old shoots have been damaged (bark splitting, browning of tissues) pruning can be limited to the removal of the damaged shoots only. Pruning will then be done starting from the top down to the point where the wood tissue is still healthy. In this way, it is possible to achieve vigorous growth and a rapid reconstitution of the canopy (Fig. 6.17). It is not advisable to spare the damaged skeleton by less severe pruning, because the plant will respond weakly and the pruning in the following years will be time-consuming. If the new shoots are formed from tissue not perfectly functional, the new growth will likely be weak and non-uniform for a few growing seasons.

In the case of severe damage, the elimination of the entire canopy is recommended. This is based on the experience following the frosts of 1956 and 1985 in Italy, when the best results to reform the canopy of damaged olive plants were obtained by cutting the trunk at the base (Sillari, 1966). In cases where the reconstitution of plants was attempted by less severe pruning, the regrowth of plants was not so vigorous, and it took more time and labour to reform the canopy. Mladar and Kovacevic (1990) showed that plants resuckering after complete elimination of the canopy grew more vigorously and yielded better than plants in which most of the old canopy was maintained while new suckers were trained to reform the whole plant.

The cut is done with a chain saw as close as possible to ground level in the early spring, but there is no need to excavate the soil around the crown. Following the removal of the old canopy, numerous suckers will emerge (when feasible, the suckers originating from the rootstock in grafted plants should be eliminated). The number of suckers will depend on the size and morphology of the section of the trunk. The longer the circumference, the more adventitious buds will break (see also sections 3.5 and 8.7). No thinning of suckers is necessary for at least three years during which the root system forms a canopy large enough to re-establish the functional balance between the above- and below-ground parts. When this happens (usually on the third year), the new canopy starts flowering and bearing fruit. Instead, if the suckers are thinned early, juvenile characteristics will be retained longer and the onset of production delayed.

The thinning of suckers will be adjusted to the final training system. For systems with multiple stems like the vasebush or bush, one can thin the suckers to 10 stems in the third or fourth year; they will be further thinned to the final three or four in the following years. It is important to choose suckers firmly inserted on the old crown as they will be less subject to break. For single-trunk systems, the thinning of suckers will have to be done early, but there is no need to select the new trunk immediately because it would constrain vegetative growth of the new canopy excessively and delay the onset of production. All suckers are allowed to develop in the first growing season, to be reduced to only 8 to 10 at the end of the second growing season. In the third or fourth growing season after resprouting, the stem most suitable to become the single trunk is identified, while the other stems are eliminated. Similar considerations are valid when a new canopy has to be formed after fire damage.

Figure 6.18 Stems of a young olive plant badly damaged by hail. To recover this plant, the canopy has to be reformed by cutting the stem at about 0.15 m above ground level and training a new stem to a vertical stick.

The bark of trees can be sunburned in areas of high solar irradiance. In this case it is better to prune less severely and maintain more foliage to protect the bark from direct solar radiation. Training systems with a dense canopy are more suitable than the vase where the central part of the canopy is relatively void of foliage.

Hail damage is usually not so dramatic as to require the suppression of the entire canopy. However, in young plants when the number and size of injuries are large (Fig. 6.18), it is necessary to cut down the plant at about 0.1–0.2 m above ground. Only the most vigorous shoot will be left to grow after resuckering and trained vertically by tying it to a stick to form the young plant as in the nursery.

If the trunk has to be preserved at all costs (see section 7.5), the best method is to cut it at the height where previous branches originated. In this case, there will be a strong vegetative response and the new suckers should be periodically eliminated to avoid competition with watersprouts.

Old trees are often damaged by wood rot, which can be caused by several species of fungi (*Coriolus* spp., *Fomes* spp., *Stereum* spp.). The surgical cure of wood to eradicate the rot should be practised in special cases only (e.g. trees of historical interest). The trunk damaged by wood rot should be cut at the base as close as possible to soil level. The wood should be excavated as far as the extension of the rot, which often expands as far as the large roots (Morettini, 1972). Once the rotten wood has been completely eliminated, the wounds can be treated with copper products at 10% in an oil- or latex-based paint. The whole surgical treatment may take several hours for a single tree.

Chapter 7

Criteria for selecting the training system

7.1 Physiology

The main physiological characteristics which are relevant for pruning have been discussed in Chapter 3. It should be evident by now that it is not difficult to meet the physiological requirements of the olive species. In brief, to achieve sustainable and abundant yields the pruner should consider:

- the vigorous vegetative response to cutting;
- the biennial reproductive cycle;
- the biological window for optimal timing;
- interception and penetration of light;
- plant age.

Yield is related to light interception, leaf area development, and water use. Bongi and Palliotti (1994) reported a low LAI for olive canopies when compared with the LAI of other fruit crops. Gucci et al. (1999) attributed the low LAI of olive plants to the non-random distribution of leaves in canopies with dense foliage.

Olive trees trained to central leader, supposedly intercepting more light per unit area of land, did not prove more productive than trees with other training systems (see sections 9.1 and 9.2). Moreover, increasing the planting density beyond 300 or 400 trees per hectare did not result in higher revenues for dryland and irrigated olive orchards, respectively (Civantos and Pastor, 1996; Tous et al., 1999; see also section 7.3). It is likely that genetic constraints in assimilate partitioning to the fruit play a greater role than light interception *per se* in present cultivars of olive.

7.2 Method of harvesting

The harvesting method is the most critical factor for the selection of the training system. The choice of the harvesting method is essentially an economic question. The most commonly used methods are shown in Fig. 7.1.

Olives are traditionally harvested manually (Fig. 7.1a). This is still the best method to harvest table olives, especially cultivars to be cured green-ripe or with a soft flesh, since cullage and bruising of the epicarp are kept to a minimum. However, the efficiency of labour in manual harvesting is low: one worker can harvest a maximum of about 10–20 kg of fruits per hour (including transport and layout of nets). The use of hand-held mechanical combs (Fig. 7.1b) increases work efficiency to over 40 kg of fruit harvested per hour per worker. This type of equipment is the best alternative to hand-picking, because it requires only a small investment and it

Figure 7.1 Different methods of harvesting olive trees: (a) manual; (b) hand-operated mechanical combs; (c) mechanical by trunk shaker; (d) mechanical by rotating combs.

can be easily adapted to all cultural conditions (small orchards, old trees, and orchards on slopes). Nevertheless, the use of mechanical combs or vibrators is tiring and proper rest periods are recommended (Vannucci and Limongelli, 1996).

The most reliable method to harvest olive trees mechanically is by trunk shakers (Fig. 7.1c). There are several types of shakers available commercially. The new-generation shakers are small and work at a high frequency of cycles and a narrow amplitude of strokes (less than 10 mm). These shakers are efficient, and more easily adaptable to single branches or small trees, and reduce the probability of bark damage. The shaker's clamp should be attached to the trunk at an angle of about 90°; otherwise the bark is more easily damaged (Martin et al., 1994b). The actual shaking may last less than 10 seconds. It can be safely assumed that between 3 and 10 minutes are needed to harvest a tree under optimal conditions (including the time needed to lay the nets underneath the tree and to recover fruits from the nets).

Yield efficiency with mechanical harvesting by shakers depends on the cultivar and ripening stage (Antognozzi et al., 1990; Famiani et al., 1998). Cultivars with large fruits or simultaneous maturation usually give higher yields. In general, most cultivars yield about 80–85% of fruits present on the tree. Famiani et al. (1998) estimated that an average of 88% of fruits could be harvested from 12 cultivars (the range was 64–97%). This figure included yield to four table cultivars, all of which yielded more than 90% of their crop. Postponing harvest by one month (from mid-November to mid-December in the northern hemisphere) increased yield efficiency from 80–89% to 89–94% in four oil cultivars (Famiani et al., 1998). Tombesi (1996) tested five cultivars in three sites in central Italy and reported an increase in yield efficiency when harvesting was postponed from mid-November to the beginning of December (note that delaying harvesting often reduces the polyphenol content of the oil). Martin et al. (1994b) and Colorio (1997) reported a more conservative figure of 60–80% of fruits shaken mechanically from cultivars grown in California and central Italy respectively. Fruits remaining on the tree need to be harvested either manually or with the shaker a second time when the fruit retention force becomes lower. In both cases there is a remarkable increase in costs.

The work efficiency of mechanical shakers ranges from 70 to 400 kg of fruits harvested per man hour. The upper limit is reached with multi-directional shakers bearing an incorporated catching frame, which unfolds beneath the tree and intercepts the fruits directly without having to lay the nets manually. Shakers with a catching frame are large, and therefore, work best when trees are not planted too close on the row and the lowermost branches are inserted at a height of at least 1.2 m from the ground.

The use of shakers implies not only a high investment for the machine, but also an orchard designed for rapid and efficient operation of the harvester. It has been estimated that about 7–10 ha planted at a density of 250–300 plants/ha yielding about 15 kg of fruit per tree is the minimum orchard size for purchasing a shaker (Famiani et al., 1998). With more productive trees, or trees planted at a higher density, this figure can probably be lowered to about 5 ha. Plant size, crop load, slope, and distance between rows should also be considered when evaluating the option of buying a shaker. In large plantations trunk shakers are essential to concentrate the period of harvesting for optimal fruit and oil quality.

Although mechanical shakers are the best technology available at present, there are inherent factors in olive biology that limit the efficiency of this harvesting method. These constraints can be summarized as follows: small fruit mass, high fruit retention force, long flexible shoots that transmit the vibration poorly, gradational maturation and ripening of fruits, large size of trees, high cullage of table olives. In addition, there are several situations where harvesters cannot be used or can only be used at high cost, such as with old trees, trees with multiple stems, steep slopes, pendulous cultivars, and thick foliage in the lower part of the canopy hiding the trunk from the operator's view.

Shakers cannot be applied to plants younger than six years old. Fontanazza (1993) reported that shakers with a light head (85–165 kg) can be used safely on trees with an age ranging from 6–7 to 40–50 years and a trunk diameter from 0.07–0.08 to 0.4–0.5 m, respectively; he also estimated that about 0.8–1 ha could be harvested in one day at a fruit retention force less than 500 g in highly productive orchards.

Colorio (1997) reported that 10–14 machine hours, and the employment of two to five people, depending on the type of shaker, were needed to harvest one hectare of trees planted at an 8 x 8 m distance.

In most cases it is safer not to apply shakers earlier than eight years after planting. The plant age at which shakers are firstly applied will obviously depend on the actual girth of the trunk which should be at least 0.1 m in diameter. For plants in sandy soils it may be better to wait until about 10 years after planting to avoid the risk of uprooting the tree. In any case, shakers will be convenient only when plants are fully productive; otherwise the efficiency of labour will be very low and uneconomical. Martin et al. (1994b) recommended not to apply shakers when plants have been recently irrigated because the loose bark is more easily damaged by the clamp. The same authors describe a number of possible occurrences of bark injury due to clamp application and report pros and cons of mechanical harvesting of olive trees.

As an alternative to shakers, trees can be harvested by mechanical combs (vibrating or spinning). These machines usually bear flexible combs mounted on a support, or a spinning comb at the end of a long arm installed on a tractor or a self-propelled unit (Fig. 7.1d). Units with vibrating combs are better suited to harvesting dense canopies. Both types of mechanical combs cause greater damage to foliage and shoots than shakers or manual harvesting. Their main advantage is the flexibility of use since they can harvest trees with multiple stems, regardless of age, size, cultivar, and slope. Ideally, 100% of the fruits present on the tree can be harvested, but this high efficiency is reached at the price of a long time of operation. The productivity is less than that obtained by shakers. Famiani et al. (1998) estimated that no more than 67% of the labour efficiency obtained by harvesting with shakers could be achieved with vibrating combs or spinning combs on 20-year-old plants of cultivar Leccino planted at 6 x 6 m and yielding 29 kg per tree. This percentage decreased to only 22% in comparison with the labour productivity of shakers carrying a catching frame for fruit interception. Colorio (1997) and Tombesi (1996) reported higher values of productivity than Famiani et al. (1998), but still substantially lower than the work efficiency reached by shakers. Because of the high initial investment (the cost of these machines is comparable to that of shakers) and high operational cost, the purchase of these units should be recommended only in those situations when shakers cannot be used efficiently.

Mechanical harvesting does not alter the qualitative characteristics of the oil, but it can have a strong effect on the quality of table olives. The cull rate due to bruising is high in table olives harvested by vibrating or spinning combs, regardless if hand- or machine-operated. Mechanical harvesting by shakers also results in high cull rates, especially for cultivars with soft flesh (Cimato, 1989). For this reason, green-ripe olives are harvested by hand. Black-ripe olives are more often harvested mechanically since the cull rate is equal to that obtained by hand-harvesting (Brighigna, 1998). At any rate, manual harvesting is the most widespread method for high-quality table olives.

7.3 Planting density

Planting distance has to be decided upon in conjunction with the harvesting method and training system. Planting distances will affect the choice of the training system and vice-versa.

The distance between rows should allow the easy movement of machinery. The larger the equipment used to manage and harvest the orchard, the wider the distance between rows. A minimum distance of 6 m is to be recommended in orchards where most practices are mechanized. In plantations where large machinery is going to be used, it is better to leave at least 7 m between rows to move equipment easily. If the distance between rows is narrow and trees are vigorous, it is likely that additional pruning will be needed every year to prevent large canopies slowing down the movement of machinery.

Trees planted too close on the row tend to grow in height rather than expand laterally. The spacing has to be adequate both on the row and between rows or mutual shading will be aggravated. The apical part of the canopy intercepts most of the light and shades the lowermost foliage with the result that fruiting will be abundant only in the upper part of the canopy. This makes higher costs for harvesting and reduces the yield efficiency of the tree. The lower part of the canopy will bear dead wood and only few leaves and fruits. Trees planted too close on the row have to be pruned periodically to control height and decrease the density of foliage in the upper part of the canopy.

Planting density should also account for the availability of moisture in the soil and soil fertility. In traditional olive growing, trees were planted at large spacings. Planting density ranged from 50 to 150 trees per hectare depending on the

Figure 7.2 Trees trained to vase in a traditional olive grove in dryland cultivation in southern Tuscany.

Figure 7.3 Young high-density planting (foreground) and mixed cropping system (background) in southern Tuscany.

precipitation regime and soil texture (Fig. 7.2). Nowadays, trees are planted at higher densities (between 250 and 420 trees per hectare) in intensive orchards (Fig. 7.3). In areas with a mean annual temperature of 15°C, rainfall of 600–800 mm and a growing season lasting about seven months, it has been estimated that at least 30–36 m^2 of land area is needed for free-canopy plants, whereas 18–25 m^2 is sufficient for central leader trees.

Although distances have to be adjusted to the specific conditions of climate, soil type, and farm structure, 4–4.5 m between trees on the row and 6 m between rows are the minimum distances one can safely use in modern irrigated olive growing. There are no advantages in the onset of fruit production, yield, or quality that justify planting densities higher than 420 trees per hectare (Civantos and Pastor, 1996). In dryland cultivation, it has been shown that maximum economic return is reached at a density of about 300 trees per hectare (Tous et al., 1999). Therefore, increasing the number of trees planted per hectare beyond these thresholds results only in higher costs at planting and for pruning in the following years. Spacings of 6 x 3 m have often been recommended for central leader systems (Fontanazza, 1993), but this density is too high and uneconomical with the current cultivars. Trees planted at narrow spacings (6 x 3 m) and trained to central leader proved more expensive than central leader trees at a 6 x 6 planting distance; the higher costs were not compensated for by the higher production (Angeli et al., 1995). The high investment in trees planted at a distance of 3 m along the row becomes even greater when trees grow, since more pruning is needed to reduce height and mutual shading. These problems are exacerbated with vigorous cultivars in fertile soils. Such recommendations on planting density will change once dwarfing rootstocks and compact cultivars, currently not available for olive, are developed.

7.4 Farm structure and social factors

Under many circumstances organizational and social factors become more important than economic aspects in determining the best pruning strategy and the most suitable training system. Some limiting factors are the absence of skilled labour, the low number of permanent workers, and the little time that the owner can dedicate to pruning in family-owned farms. Training systems allowing more flexibility in the pruning strategy are more easily adapted to specific farm organization. In large orchards, pruning all the trees may take too long and be incompatible with labour available locally. In cases when permanent labour is employed, but the size of the olive orchard makes it impossible to prune all plants in one year, it is better to prune only part of the orchard every year (see also section 4.7). On the other hand, if pruning is performed by hiring temporary labour it may be more practical to have all the trees pruned in the same year every two or three years.

If only part of the orchard is pruned every year, the time needed for pruning the whole orchard can be divided by the number of years between two subsequent prunings, so that each year only a share of the plants are pruned. For example, if 115 hours per hectare are needed to prune all the plants every other year, only 58 h/ha will be needed to prune half of the plants every year, and 46 h/ha to prune one-third of the plants every year (data taken from Table 4.3; note the remarkable saving with biennial pruning even though the pruning of individual trees takes longer).

Using these figures for a 20 ha orchard (density of 333 trees per ha) on a farm where only two pruners are permanently employed, 1665 man hours (208 days) are needed to prune the whole orchard. The two pruners will take 104 days, too many to complete pruning within the optimal biological period. By pruning half of the orchard every year, only 52 days will be needed per worker. If a triennial interval is chosen, 35 days per pruner are needed for pruning only one-third every year, provided that conditions are suitable to sustain such a long interval (see section 4.6). Plantations managed by pruning only part of the whole area show smaller fluctuations in yield due to alternate bearing, because at least some trees are out of phase from the rest of the orchard.

7.5 Aesthetics

The knotty trunk, the expanded canopy and the silver-green colour of foliage make the olive species suitable to be used as an ornamental plant. The fruitless cultivar Swan Hill is widely grown in Arizona as an ornamental because of its abundant flowering and small output of airborne pollen (O'Rourke and Buchmann, 1986; Sutter, 1994). Individual trees can be found in city streets, gardens and parks (Figs 7.4; 7.5). The olive plant is also suitable for forming hedges with thick canopies almost impenetrable to sight. In arid and semi-arid zones the use of olive plants in private gardens and public parks is becoming more common because of the exceptional drought resistance of this species. In many cases trees have grown so

Figure 7.4 Olive tree in the historical Alfama district of Lisbon.

old and large that they represent territorial landmarks and a cultural reference for local communities. Because of its beauty and longevity the olive tree is an essential part of the landscape of many Mediterranean agricultural areas. Besides the more favourable market trends for olive products in recent years, olive growing has also been revived in some areas in consideration of the aesthetic value that this crop adds to the farm and landscape.

The pruning of olive plants grown as ornamentals is based mainly on aesthetic criteria. Hence, the concepts and techniques discussed in the previous chapters would need to be revised. Most shoots of plants grown in pots are pinched or headed once or twice a year to obtain a well-rounded canopy, small size and bushy appearance. If olive plants are used as hedges, pruning is done quickly by hedging and topping cuts at the beginning of spring and during the growing season to increase thickness of the canopy and avoid fruiting.

When pruning monumental trees, every effort should be made to preserve the health and structure of the tree. Pruning should cure the diseased parts (see also section 6.9), remove the dead wood or parts that are structurally weak and may represent a potential danger for the public, and maintain the pleasant appearance of

Figure 7.5 A former olive grove, currently a park, in Adelaide, South Australia.

the plant. A few cuts in the upper part of the canopy can be performed to improve light penetration and allow some renewal of fruiting.

Despite the obvious cases when olive is grown for ornamental purposes, the role of aesthetics in the field should not be underrated. The psychological impact of a disordered canopy versus one that has been pruned carefully and regularly can often be more important that technical components in the selection process of the training system. A well-structured canopy architecture has certainly a better visual impact than free-canopy systems (see the box on pages 131–2 in Chapter 9). For many people "what looks good is good", but one should be careful when applying this principle to olive growing: cultivation costs are likely to increase. It goes without saying that the final decision about the training system is up to the grower's preference.

7.6 Revenue

The olive plant can be trained to many different shapes. Training systems differ in the number of trunks and primary branches, shape of the skeleton, the more or less regular canopy, tree height, and pruning techniques. Some systems require annual pruning to maintain a regular shape, usually performed without significantly altering the permanent skeleton formed during the training period. Other systems are characterised by a free canopy, which is renewed with few cuts, not necessarily done every year. The performance of training systems should be evaluated based on economics. Modern training systems should be formed and maintained at minimum costs, and produce high yields. Under most circumstances high revenues depend on the ability to reduce costs for pruning and harvesting. Since economic and social conditions may vary considerably from area to area, training systems and pruning methods are subject to be adjusted to the specific conditions of the site or farm. The main training systems used in modern olive growing are described in the next chapter.

Quick reference for selecting training systems

Many aspects need to be considered when selecting the training system. The most important ones have been discussed in the present chapter. A simplified scheme to be used as a quick reference is illustrated in Fig. 7.6.

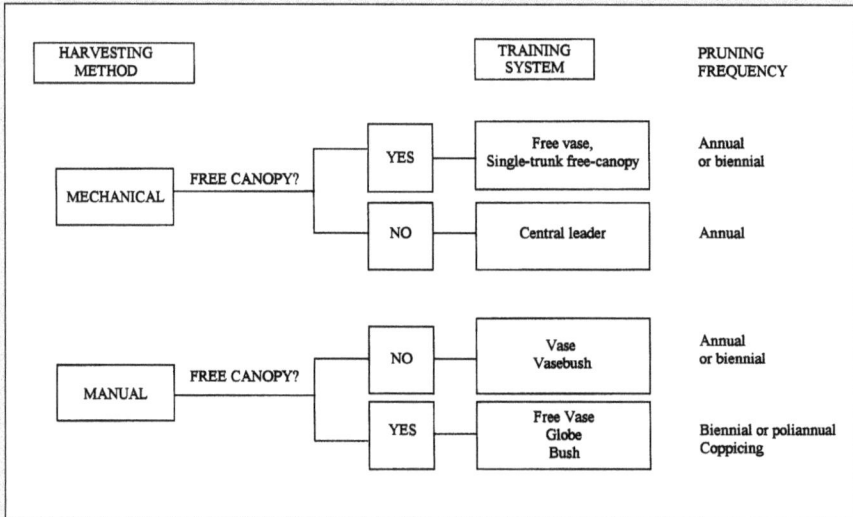

Figure 7.6 Flow chart for the selection of the training system (proceed from left to right). The frequency for pruning mature plants is also reported.

If the plants are going to be harvested mechanically by shakers, a single trunk must be formed. When choosing the training system, it is crucial to use pruning strategies that can keep pruning costs low. This is because the time to shake the tree, and the yield from mechanical harvesting, are similar in trees trained to either central leader, vase, or single-trunk free-canopy systems. Substantial savings can be obtained by adjusting the intensity and/or frequency of pruning to the specific orchard conditions. In general, this is possible with most training systems, but free-canopy systems allow more flexibility in pruning. Current trends are to prune the canopy as little as possible to keep costs low while exploiting the full fruiting potential of the tree.

Every training system is suitable if harvesting is done manually or by hand-operated equipment. However, systems restricting plant height are more economical in the long run, because most operations can be performed from the ground. The vase and the vasebush require more pruning than free-canopy systems. Free-canopy systems are also more easily adaptable to biennial pruning. A cost-effective solution is to leave the plants unpruned until the fifth to seventh year after planting, and then choose the training system between the following options: free vase, vasebush, bush (sections 6.3; 8.6; Figs 6.5; 6.6). Systems with a full canopy (globe, bush) are better suited to environments with high irradiance. Free-canopy systems yield well, but sometimes workers do not like to pick fruits from dense canopies. Open canopy systems allow more lateral expansion than central leader systems and grow less rapidly in height.

Chapter 8

Description of modern training systems

8.1 Vase

The vase has evolved from ancient times and it is probably the most common training system in many olive-growing areas. The vase has a greater surface-to-volume ratio per tree than systems with a full canopy (e.g. globe, bush). Light penetrates the interior of the canopy and it is quite evenly distributed. There are several types of vase, differing in the number of branches, their inclination, the height of the trunk, and the shape of the tree. The shapes of traditional vase systems include the cone, the inverted cone, the cylinder, a cylinder on top of an inverted cone, or multiple cones (Figs 6.12; 6.15; 8.1a, b). It is virtually impossible to describe all the different vase systems, many of them widespread only in small areas. Traditional vase systems are well described in Garcia-Ortiz et al. (1997) and Morettini (1972), and are not reviewed here.

The skeleton of the vase is formed by a single trunk of variable height and primary branches, oriented in different directions, surrounding an open space in the central part of the canopy (Fig. 6.7; 8.1a, b). The number of branches, usually inclined at 45°, ranges from three to five (Fig. 8.1). In the cylindrical vase and the polyconic vase the branches are inclined at 45° in their proximal part, but trained almost vertically at a certain distance from the trunk (Figs 1.5; 8.1). Most of the fruiting shoots are borne on secondary and tertiary branches (Figs 6.7; 8.2).

The height of the trunk depends mainly on the harvesting method. If trees are hand-harvested, the trunk can be only 0.3–0.4 m in height (Fig. 8.2) and the tree becomes similar to the vasebush (see section 8.3). For mechanical harvesting, the trunk should be at least 1–1.2 m high and free of lateral shoots. To improve yield to mechanical harvesting, the primary branches should be trained more erect and with shorter secondary branches than in hand-harvested vase systems. The more inclined these branches, the more difficult is mechanical harvesting.

The length of secondary branches is usually inversely proportional to the number of primary branches. Secondary branches are inserted mainly on the external side of primary branches, especially if harvesting is by hand; few small branches are present on the internal side of each primary branch (Fig. 8.1b, c).

The pruning of young plants is directed to form the trunk and primary branches. The tree is usually not pruned at planting or in the first growing season, except to reduce the vigour of dominant lateral shoots and watersprouts. Vigorous shoots and watersprouts are weakened by heading, or completely eliminated if excessively

Vase

Features
- Single-trunk
- Primary branches inserted on main trunk
- Three to five primary branches
- "Window" in the central part of the canopy
- Regular arrangement of secondary branches

Advantages
- Uniform light distribution within the canopy
- Suitable for different growth habits
- Suitable for table olives
- Suitable for mechanical harvesting provided that branches are kept relatively short and a rigid structure is formed

Disadvantages
- Skilled labour for pruning
- Pruning is time-consuming, especially for traditional vase systems

Figure 8.1 (a) Mature olive tree trained to free vase; (b) young tree trained to vase—the short trunk makes this tree unsuitable for mechanical harvesting; (c) Terminal of a primary branch of a tree trained to vase.

Figure 8.3 Young tree trained to free polyconic vase. The arrangement of fruiting branches, the pruning of the arrow, and thinning of shoots next to the leader of each branch are less accurate than in the traditional polyconic vase.

Figure 8.2 (a) Tree trained to free vase in the third growing season after planting. The pruning consisted of removing few lateral shoots from the trunk and the central part of the canopy; (b) low-trunk free vase in the fourth growing season after planting. The branches were selected among those developing naturally.

vigorous. The main axis is headed if it does not have good lateral shoots to be used as primary branches. Once the first truss of lateral shoots is formed at about 1–1.2 m in height, the main axis is headed. Heading of lateral shoots is used to widen the angle of inclination of the future primary branches. Trainers for the inclination of primary branches and bending of the lateral shoots growing below the point of insertion of primary branches are not used any longer because it is too expensive. The choice of primary branches can be postponed until the third or fourth year after planting; during this period pruning should favour the growth of selected shoots without eliminating too much foliage.

The pruning of mature trees should favour light penetration in the lower part of the canopy by eliminating more shoots in the upper and central zones of the canopy. The height of primary branches should be reduced and secondary branches should be shortened periodically with heading-back cuts. The vase is usually pruned every year, but it can be easily adapted to biennial pruning.

Today, less attention is given to the formation and distribution of secondary and tertiary branches than in the past. Secondary branches should be inserted on primary branches with a relatively wide angle. In modern vase systems the final shape is also less regular than in traditional vase. To train a plant to free vase, the first thinning of shoots in the internal part of the canopy can be delayed until the fifth to seventh growing season (Fig. 8.2; see Chapter 4).

The polyconic vase, still the most common system in many old groves, is formed by four primary branches, each pruned to a conical shape with a distinct terminal end (see also Chapter 1; Fig. 1.3). The shoots growing near the apex of each branch should be adequately thinned to reduce competition with the leader (Fig. 1.5). The polyconic vase is no longer recommended for new plantings because of high pruning costs. Free polyconic vase can be obtained by reducing the thinning of shoots growing near the arrow of each branch and making fewer large cuts rather than many small ones (Fig. 8.3).

8.2 Globe

The globe (or globe vase) is a system with a single-trunk and a full canopy. The main difference from the vase is that the centre of the canopy is occupied by either secondary branches or the terminal part of the main axis. The final shape of the globe is round, hemispherical, or ellipsoidal, depending on the growth habit of the cultivar (Figs 8.4; 8.5).

The globe is widely used in areas where plants grow vigorously and sunlight is high. The high density of foliage protects the bark from direct solar beams. Trees trained to globe are usually pruned every year, but they can be easily adapted to longer intervals of pruning (two to four years), especially under irrigated conditions or climates with a short summer drought. Note that the fruiting surface of trees

Figure 8.4 Young trees trained to globe in a high-density irrigated orchard. No pruning is done in the first three years after planting.

Globe

Features
- Single trunk of at least 1.0–1.2 m in height
- Primary branches as in the vase
- No "window" in the central part of the canopy

Advantages
- Suitable for mechanical harvesting by trunk shakers (see vase)
- Suitable for environments with high light levels
- Suitable for different growth habits

Disadvantages
- Excessive density of foliage if pruning is neglected
- Skilled labour for pruning

Figure 8.5 Old trees trained to globe in a traditional dryland orchard.

pruned lightly or every three or four years becomes confined to the external part of the canopy after a few growing seasons, and productivity declines. Hence, the pruning should open sufficiently large gaps for light penetration, air circulation, and renewal of fruiting. The minimum recommended distance on the row between trees trained to globe is 4.5 m. The actual spacing will obviously depend on water availability and other cultural factors (see section 7.3).

The pruning of young plants is similar to that described for the vase with regard to the formation of the single trunk and selection of the primary branches (Fig. 8.4). The main axis is headed in the second or third year after planting to favour development of the lateral structures; however, today the axis is often left, to be removed later or not removed at all, because the lateral shoots usually overgrow it. In this case the terminal part of the main axis fills the centre of the canopy. Light thinning of shoots may be necessary to stimulate the growth of shoots that will form the skeleton of the tree. The angle of inclination of upper secondary branches is relatively narrow to avoid excessive shading of branches beneath. Erect, relatively short branches are preferred for trees to be mechanically harvested by shakers.

The pruning of mature trees involves the periodic shortening of the primary branches to prevent their excessive elongation and consequent lateral expansion of the canopy. Nowadays, to reduce costs, renewal of the fruiting surface is obtained by eliminating entire secondary branches rather than thinning single fruiting shoots (section 6.2). Only few cuts are done every year to improve light penetration. Large heading cuts may be necessary every two to four years to open windows and give the canopy a more globular shape.

Vasebush

Features
- Primary branches originating from the soil line or inserted on a short trunk
- Secondary branches as in the vase

Advantages
- Limited height
- Early onset of production
- Suitable for different growth habits
- Suitable for table olives
- Easily formed from suckers after cutting the plant at ground level

Disadvantages
- Unsuitable for mechanical harvesting by trunk shakers

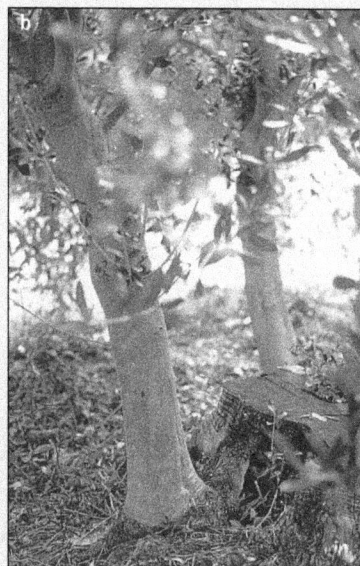

Figure 8.6 (a) Plants trained to vasebush with the canopy reconstituted after the great frost of 1985 in Italy; (b) close-up view of suckers that sprouted after cutting the main trunk at ground level—these suckers are now forming the primary branches of trees trained to vasebush; (c) mature tree trained to vasebush.

8.3 Vasebush

The vasebush is a vase without a proper trunk and with primary branches originating directly from the crown of the tree (Fig. 8.7) or inserted on a very short trunk (Figs 4.9; 8.7; 8.15). It became widely used in the late 1950s to reconstitute orchards damaged by frosts. The development of the vasebush also represented the first attempt to adopt a less labour-intensive system than the polyconic vase or other traditional systems. The number of branches of the vasebush varies, but typically ranges between three and six. The main advantage of this training system is that the plants are kept shorter than if trained to vase or globe, so that most operations can be performed from the ground. However, because of the lack of an appropriate trunk for attaching a shaker, the vasebush can be recommended only when fruits are harvested by hand or hand-operated equipment. It is an excellent system for table olives.

Figure 8.7 Plant trained to vasebush before (a) and after (b) pruning in the fourth year after planting. Pruning consisted of few cuts in the upper central part of the canopy to improve light penetration.

The canopy of the vasebush is allowed to grow more or less freely, depending on the method and severity of pruning. In the original design, the primary branches were trained similarly to those of the polyconic vase (Fig. 8.6a, c), but pruned more heavily towards the central part of the canopy. Attempts to form the vasebush by planting three individual plants, each representing a single branch, proved more expensive and less productive than the planting of a single tree with three or four branches (Garcia-Ortiz et al., 1997; Morettini, 1972). Nowadays, to reduce pruning time and costs, the structure of secondary and tertiary branches is not formed so regularly as in the past (Fig. 8.7).

During the training period, pruning is aimed at forming the primary branches as described for the vase. Heading cuts may be necessary to stimulate the emergence of vigorous lateral shoots in the lower part of the stem of the young plant. As the plant grows, lateral shoots are thinned to favour the growth of the main axis of each branch and reduce competition between shoots that will form the secondary branches. Watersprouts growing in the inner part of the canopy and suckers should be removed. The vasebush can also be obtained from plants left almost unpruned until the fifth to seventh year after planting, when few cuts are made in the central part of the canopy for light penetration (see section 8.6; Figs 8.7; 8.15). If the vasebush is obtained from a plant that has been cut at ground level (Fig. 8.6), 6 to 10 of the most vigorous and well-spaced suckers are allowed to grow for the first three or four years (see also section 6.9).

The guidelines for the pruning of mature trees are similar to those described for the vase. The pruning of mature plants is preferably done by elimination of entire secondary fruiting branches rather than single shoots.

8.4 Central leader

The central leader (also termed monocone or monocone vase) was developed in the 1930s as an alternative to the polyconic vase for high-density plantings (Roventini, 1936), but it has become a common training system only in the last 15 years.

The central leader consists of a single trunk, free of any lateral shoot up to 1.0–1.2 m in height. Primary branches are numerous and arranged in an elicoidal fashion along the central axis for maximum occupation of space and minimum overlapping. The primary branches are of decreasing length from the base to the top of the tree. The tree assumes a conical shape with a distinct terminal (Fig. 8.8), similar to individual branches of the polyconic vase.

The branches are selected with a wide crotch angle (Figs 8.8c; 8.9); shoots and branches inserted with a narrow angle tend to grow vertically, making the inner part of the canopy dense and competing with branches located above as well as the central leader itself. Overlapping branches should be spaced by at least 1 m along the central axis. The skeleton structure must be as rigid as possible, with branches no longer than 2–2.5 m to transmit the vibration efficiently from the shaker to the fruit-bearing shoots.

The training of the young plant is aimed at eliminating the lowermost lateral shoots gradually, so that 1.0–1.2 m of the single trunk is free of any ramification by the end of the fifth or sixth year after planting. The shaker cannot be applied earlier than the sixth growing season after planting. The most vigorous shoots and those growing in the lower part of the trunk are removed first because they have a greater probability to overgrow the branches that will form the permanent structure of the tree (see also section 5.5).

Lateral shoots inserted along the main axis need to be thinned to favour the growth of the central leader as well as that of the shoots that will form the lateral scaffolds

Central leader

Features
- Single trunk of at least 1.0–1.2 m in height
- Branches of decreasing length, elicoidally arranged from the base to the top of the main axis
- Conical shape

Advantages
- Suitable for mechanical harvesting by trunk shakers
- Homogeneous appearance of trees in the orchard
- Suitable for table olives

Disadvantages
- High pruning costs starting from planting
- Skilled labour for pruning
- Excessive height of trees
- Onset of production, yield, and oil quality not better than other modern systems
- Unsuitable for pendulous cultivars

Figure 8.8 (a) Mature tree (cv. Leccino) trained to central leader. The picture was taken four weeks after pruning; (b) rows of central leader trees planted at a 6 x 5 m distance in an irrigated orchard. Note the conical shape of trees; (c) young tree trained to central leader in a 6 x 3 m planting. Note the even distribution and wide crotch angle of shoots that will become the primary branches of the mature tree.

Figure 8.9 Detail of the canopy of a central leader tree. Note the heading-back cuts performed on a primary branch to renew fruiting, maintain a wide angle of inclination, and contain the lateral expansion of the tree.

of the tree. Lateral shoots competing with the central leader can either be removed entirely or headed. Heading is preferably used to weaken a lateral shoot without reducing the plant leaf area excessively, or to widen the crotch angle of the lateral structure (Figs 5.5; 5.6). The removal of lateral shoots and watersprouts is started from the first growing season after planting and it is performed before the end of the summer. Particular attention has to be paid to the thinning of lateral shoots competing with the central leader (Fig. 5.7). One of the two opposite shoots is usually eliminated on each node of the main axis (Fig. 5.5). Overlapping or overcrossing shoots, and watersprouts inserted at a narrow angle, are also removed. The leader is tied to a trainer to maintain the vertical position; the trainer should be about 1.8 m above ground and will be removed once the tree structure is developed, which usually happens in the fifth or sixth year after planting.

The pruning of mature trees will be aimed at reducing canopy height, containing lateral expansion of branches, renewing fruiting shoots, and eliminating watersprouts and suckers. Once the tree grows over 4.5 m the central leader is shortened by about 1 m next to an upright lateral shoot that will form the new arrow. The leader has to be replaced every two to four years, depending on the vigour of the tree, cultivar, soil fertility and soil water availability. The terminal part of primary branches is headed to reduce lateral expansion of the canopy, renew the fruiting surface, and maintain the conical shape (Fig. 8.9). Trees should be spaced at least 4 m on the row, because trees grow too high and mutual shading becomes excessive at narrower distances (Fig. 8.10).

The central leader system has been promoted in recent years as the best system for highly mechanized plantings. However, the pruning is quite elaborate and expensive starting from the training phase, which further increases the capital invested before there is any return. The average time for pruning central leader trees was 47 h/ha per year in 12 different trials in Tuscany (Table 8.1). Only in one case was production started early, in the third year after planting (Table 8.1). The few published works indicate that the pruning needed to form and maintain trees to central leader takes longer than that required for other modern training systems. The theoretical advantages of the central leader (early onset of production, high production, high yield to mechanical harvesting) have not been confirmed by studies where the central leader was compared with other training systems (see also Chapter 9). The conical shape of the tree is also more difficult to maintain than other shapes that more

Figure 8.10 Central leader trees (cv. Frantoio) planted at 6 x 3 m under irrigated conditions in the 11th year after planting. These trees are over 6 m high.

Figure 8.11 (a) Central leader tree during the growing season. Low-hanging shoots need to be removed to facilitate cultural practices and mechanical harvesting; (b) disordered canopy of the semi-pendulous cultivar Frantoio trained to central leader in the same orchard as in (a).

Table 8.1 Performance of olive trees trained to a central leader at different sites in Tuscany. Onset of production estimated when fruit yield exceeded 0.15 kg per plant. Source of raw data: ARSIA (1999).

Planting distance (m x m)	Irrigation	Onset of production (years after planting)	Yield (kg fruit/ plant)	(t/ha)	Pruning cost (% annual cultivation cost)	Pruning time (h/ha)	Years of production
6 x 3	Yes	N.A.	1.7	1.0	42	N.A.	5
6 x 3	Yes	4	5.0	2.77	21	46	10
6 x 3	Yes	4	11.5	6.38	24	107	10
6 x 3	Yes	3	4.9	2.72	N.A.	74	8
6 x 4*	Yes	4	2.6	1.09	23	28	8
6 x 6	Yes	4	7.8	2.16	40	113	7
6 x 6	Yes	5	7.9	2.19	20	52	10
6 x 5	No	6	3.0	1.02	26	15	7
6 x 6	No	5	7.1	1.97	15	30	7
6 x 6	No	4	9.2	2.56	20	21	10
6 x 6^	No	4	1.6	0.44	N.A.	N.A.	5
6 x 6	No	5	5.0	1.38	37	38	9
6 x 6	No	6	3.7	1.03	23	16	5
6 x 6	No	4	7.4	2.05	N.A.	26	7

N.A. Not applicable;
^ Frost damage;
* Free central leader.

closely resemble the natural growth habit of the species. This is particularly evident with pendulous and semi-pendulous cultivars (Fig. 8.11). Some problems also may arise with vigorous cultivars which tend to produce many watersprouts in response to the pruning during the training phase.

Therefore, the central leader is no more convenient than other modern training systems and should not be recommended for hand-harvested orchards. To keep pruning costs as low as possible the original guidelines on the inclination, length, and crotch angle of branches and conical shape of the central leader should not be followed. In this way, the shape of trees may look more similar to single-trunk free-canopy systems than central leader proper. In conclusion, the central leader is only one of the possible alternative systems that can be used in modern olive growing.

8.5 Single-trunk free-canopy (STFC)

Free-canopy systems include all those requiring little or no pruning (see also sections 8.6 and 8.7). Although beneficial effects of reduced pruning have been reported for centuries, free-canopy systems are becoming more common only recently, thanks also to the more widespread use and optimization of means other than pruning to control plant vigour. The main advantage of free-canopy systems is the short time needed to prune a plant without detrimental effects on the onset of production or yield. Pruning can be adapted to the specific needs of the individual plant as assessed by vigour, crop load, size, and growth habit.

Single-trunk free-canopy (STFC) systems combine the feature of a single trunk with the low cost of and flexibility in pruning. The STFC is similar to the globe, but primary branches are not necessarily regularly distributed along the main axis and minimum pruning concepts are applied to the whole canopy. As a result, the skeleton of the STFC does not correspond to any predefined, regular structure, but tends to adapt to the natural growth habit of the cultivar (Fig. 8.13).

Figure 8.12 Young tree trained to single-trunk free-canopy in the second growing season after planting. The tree is growing unpruned, except for few lateral shoots removed from the lowermost part of the trunk.

In brief, STFC systems are characterized by:

1 a single trunk of at least 1.0–1.2 m in height (Fig. 8.12) for easy attachment of the shaker's clamp, obtained as previously described (sections 5.5; 7.2; 8.4);

2 a canopy not rigidly shaped, but allowed to grow as free as possible (Fig. 8.14). A number of primary branches are selected among those developing naturally, depending on their spacing along the trunk (at least 0.1 m), orientation for good light interception, and their relative vigour. Since no predefined shape of the canopy is pursued, STFC trees may appear non-homogeneous;

3 pruning should be limited to the removal of a few shoots to avoid crowding and excessive shading inside the canopy, elimination of suckers and the most

Single-trunk free-canopy (STFC)

Features

- Single-trunk of at least 1.0–1.2 m in height
- Lateral branches selected among those developing naturally on the main axis
- Free canopy

Advantages

- Low pruning costs
- Suitable for mechanical harvesting by trunk shakers
- Suitable for different growth habits

Disadvantages

- Non-homogeneous appearance of trees in the orchard
- Experience required to determine the intensity and frequency of pruning

Figure 8.13 (a) Tree trained to single-trunk free-canopy (STFC) in an irrigated orchard; (b) trees trained to STFC at a 5.5 x 4.5 m planting distance in an irrigated orchard at the end of the third growing season; (c) trees same as in (b) at the end of the sixth growing season.

Figure 8.14 Trees trained to single-trunk free-canopy (STFC) at 5.5 x 4.5 m spacings in an irrigated orchard.

vigorous watersprouts, and balancing of branches of different vigour. Pruning of very low-hanging shoots is useful to ease cultural practices on the row. The renewal of fruiting shoots is done preferably by suppressing entire fruiting branches and replacing them with young vegetative shoots properly located in the canopy (sections 6.2; 8.1).

STFC systems are the result of the application of minimum pruning concepts to satisfy the needs of modern olive growing. Pruning should be limited to less than five minutes per plant during training and 15 minutes for a mature tree (Table 9.2). Pruning costs for STFC systems are less in comparison with other systems suitable for mechanical harvesting, provided that the pruner is able to interpret the plant status and prune accordingly. Ideally, the pruning of STFC trees requires no skills, but in reality, experience is needed since the pruning technique has to be adjusted depending on tree growth and shape. However, after sufficient practice in visually assessing the plant status (section 4.5), the pruning of STFC can be performed quickly by unskilled labour.

8.6 Bush

The bush (or free bush) is a free-canopy system which is obtained with minimum pruning from the time it is planted. The canopy is allowed to grow as free as possible so that the final shape resembles that of naturally growing plants (section 3.3; Figs 3.6; 5.8; 5.9). The trunk is absent or short, with lateral scaffolds that make it unsuitable for harvesting by shakers. The bush is also used for plants grown virtually without pruning after resprouting (Fig. 8.17; section 8.7).

Bush

Features
- Free canopy

Advantages
- Hardly any pruning during training
- Minimum pruning of mature trees
- Unskilled labour for pruning
- Early onset of production
- Suitable for different growth habits
- Easily convertible to vasebush or free vase

Disadvantages
- Unsuitable for mechanical harvesting by trunk shakers
- Non-homogeneous appearance of plants in the orchard

Figure 8.15 Twelve-year-old plants (cv. Leccino) in a non-irrigated orchard at 6 x 6 m spacing. Plants were grown unpruned for the first five years after planting. In the spring of the sixth growing season a light pruning was performed to eliminate three or four branches in the central part of the canopy. Then, plants were not pruned for another six years. (a) Before pruning; (b) after pruning; (c) after pruning to vase whereby only four to six major branches were selected and gaps were opened in the central part of the canopy.

Figure 8.16 First pruning of an olive plant trained to bush at the beginning of the sixth growing season after planting. The pruning does not alter the natural shape of the plant, but improves light penetration in the central part of the canopy.

There is hardly any pruning done during the training phase. Pruning is limited to the elimination of suckers and vigorous watersprouts when needed. If the canopy is well balanced and shoot growth rate is adequate, the plant does not need any pruning for five to seven years after planting (section 5.6; Figs 5.8; 5.9). Then, one to four branches (stems) are removed from the central part of the canopy to improve light penetration (Figs 8.15; 8.16). In this way, plants can grow practically undisturbed for another four to six years, depending on water availability, soil fertility, and length of the growing season. Then, heavy pruning is needed to reform the canopy. There are several options available. Plants can be gradually converted into other systems (vasebush, globe), maintained as a bush, or cut at ground level to start a coppicing cycle (Figs 5.8; 8.15; 8.17).

To be converted to globe, plants will be pruned once shading becomes excessive and annual shoot growth weak, which usually happens after a few cropping years (8th to the 10th year after planting). The pruning consists in the selection of major branches among those already developed and few severe cuts to increase light penetration, remove dead wood, and contain canopy volume. In the next two years the secondary branches are also selected. To convert a bush into a vase the pruning should remove branches and shoots to open a large window in the central part of the canopy (Figs 5.8; 8.1b; 8.15). If the plant is to be maintained as bush, plants are pruned more severely the first time with gaps opened in the canopy by few major cuts (Figs 8.15; 8.16), and then lightly in the following year. The alternation of severe and light pruning can be repeated for a number of years, or else biennial pruning can be adopted under most circumstances. Every few years the length of major branches should be reduced by heading cuts. The renewal of fruiting shoots is obtained by few large cuts.

Both pruning and harvesting are conducted from the ground since plants do not grow in height as much as in other training systems. Because of lateral expansion in volume, plants grown to a free canopy have to be spaced at least 5 m on the row. Elimination of entire branches is preferred to heading to improve light penetration and contain plant size. The pruning of low-hanging shoots is useful to facilitate soil cultivation as well as laying of nets and recovery of fruits at harvest.

Table 8.2 Performance of olive trees trained to a free canopy (bush) with minimum pruning at different sites in Tuscany. Onset of production estimated when fruit yield exceeded 0.15 kg per plant. Source of raw data: ARSIA (1999).

Planting distance (m x m)	Irrigation	Onset of production (years after planting)	Yield (kg fruit/ plant)	(t/ha)	Pruning cost (% annual cultivation cost)	Pruning time (h/ha)	Years of production
6 x 5	Yes	4	13.1	4.36	8	20	9
6 x 5	Yes	3	2.4	0.8	11	6	6
6 x 5	Yes	4	11.6	3.86	8	8	8
6 x 5	Yes	4	5.7	1.9	20*	12	7
6 x 6	Yes	5	6.9	1.94	5	N.A.	5
6 x 5	No	3	9.4	3.13	11	19	7
6 x 5	No	4	11.1	3.71	6	3.5	9
6 x 5	No	4	3.9	1.31	19	17	7
6 x 6	No	4	6.6	1.83	34	42	8
6 x 6^	No	2	8.2	2.27	N.A.	N.A.	5
6 x 6	No	4	6.7	1.87	5	4	5
6 x 6	No	3	10.4	2.88	N.A.	5	7

*Includes soil management;
N.A: Not applicable;
^ Frost damage

The main advantages of the bush system are the minimum cost of pruning, the early onset of production, and the high productivity beginning from the onset of production (Table 8.2). Pruning is easy and can be performed by unskilled labour. An annual pruning time of 6 h/ha is not difficult to achieve in the first 10 years after planting. Therefore, the bush allows the grower to save considerably in costs and investment of plants, while productivity is comparable to that of other systems.

The biggest problem is the unsuitability to harvest by shakers, which, at present, limits the utilization of the bush to orchards where fruits are harvested manually. In addition, although to the best of our knowledge there are no trials where the bush has been tested for table cultivars, it is likely that the pruning should be modified for adequate fruit thinning and better light penetration in the canopy.

8.7 The coppiced orchard

An alternative method of managing the olive orchard was developed by Balilla Sillari 20 years ago to minimize pruning costs and simplify orchard management for oil production (Cantini et al., 1998). To manage the olive orchard by coppicing, the method consists in exploiting the high resprouting capacity of this species after cutting at ground level. This system has been tested at one site in southern Tuscany over 19 growing seasons and has been applied to four other different trials since 1995. The method and results from the main experiment are reported here.

The original experiment was carried out in a 6 x 6 m orchard managed by coppicing beginning in spring 1980, when it was divided into ten plots of about 25 plants each and all plants in one plot were cut off at ground level (approximately 5–10 cm from the point of emergence) with a chain saw (Fig. 8.17a, b). In the following years coppicing was extended to one additional plot per year until 1989. Every year the plot to be coppiced was selected randomly from those plots of the initial ten that had not yet been coppiced. In this way, all plots had been coppiced once at the end of the ten-year cycle. In March 1990 the first plot was coppiced again and a second cycle of coppicing was started. As a result of coppicing, the orchard consisted of unevenly-aged plots. The performance of the coppiced system was compared with that of an adjacent orchard where plants were trained to vasebush and pruned yearly by selective elimination of about 25% of the fruiting shoots.

Coppicing was done no earlier than budbreak to minimize the risks of damage by spring frosts, which occasionally occur in the study area. Coppicing may be performed earlier in warmer zones, where frosts are less frequent. In the coppiced system all the suckers from the stump are allowed to grow freely without any pruning until the third year after coppicing, when four to ten suckers, located in the central part of the canopy, are eliminated to increase light penetration (Fig. 8.17c, d; section 3.3).

Figure 8.17 (a) General view of the coppiced olive orchard. The stumps remaining after the coppicing of trees are shown in the foreground. Nine-year-old trees, which will be coppiced the following year, are in the background; (b) Close-up view of the stump immediately after coppicing, showing ten-year-old cuts (arrows) and more recent ones. Bar equals 50 mm; (c) Olive plants at the end of the first growing season after coppicing. Suckers were allowed to grow freely without any pruning; (d) Plants at the end of the third growing season after coppicing, when a few suckers are eliminated in the central part of the canopy. The prunings are shown in the foreground; (e) Eight-year-old trees in full production in the coppiced olive orchard; (f) Fruiting shoots on a nine-year-old tree growing in the coppiced orchard (reprinted with permission from Cantini et al., 1998, *HortTechnology* 8: 409–12).

The coppiced orchard

Features
- Free canopy
- Plants cut at ground level every 8–12 years
- Pruning takes only 14 min per plant over 10 years

Advantages
- Minimum cost for pruning
- Unskilled labour for pruning
- Early onset of production
- Yield and oil quality comparable to the vasebush
- Reduction of alternate bearing
- Control of plant size
- Suitable for different growth habits
- Rejuvenation of old trees
- Low cost implementation

Disadvantages
- Unsuitable for mechanical harvesting by shakers
- Non-homogeneous appearance of plants in the orchard
- Unsuitable for table olives
- Limited experimental basis

The time required for coppicing is about 8 minutes per plant, whereas that for the thinning of suckers is 6 minutes (Table 8.3). No pruning was done in the first and second year after coppicing, or from the fourth to the tenth year. Therefore, the coppiced system required 14 minutes per plant over the ten-year cycle, or 6.5 h per ha per year (Table 8.3). This value is much lower than the time needed to prune other training systems.

Table 8.3 Summary of pruning practices and time required to manage the olive orchard by coppicing during a ten-year cycle. Time per hectare was calculated considering the planting density of 277 plants/ha (reprinted with permission from Cantini et al., 1998, *HortTechnology* 8: 409–12).

Year	Pruning practice	Labour (min/plant)	Labour (min/ha)
0	Coppicing	8	222
3	Thinning of suckers	6	166
Total per year		14	388
Total of 10 years		14	3878

The onset of fruit production occurred in the third year after coppicing; production per plant increased up to the seventh or eighth year, when it started to decline. The fruit-bearing habit of shoots and trees was similar to that observed for other training systems (Fig. 8.17e; f). The coppiced orchard produced 91% of the fruit yield of the control in the 1989–98 period and oil yield was unaffected by coppicing. These yields were typical of non-irrigated olive trees growing in the same area and were above the average yields of olive groves in Tuscany. Coppicing also reduced alternate bearing in comparison to the vasebush used as the control (Cantini et al., 1998). There were no major effects of coppicing on oil quality, nutritional status of the tree, disease and pest control, patterns of vegetative growth and fruit development. Tree size and yield per tree were relatively uniform within each plot of the coppiced orchard.

The coppicing system can be adapted to most cultural situations and does not require changes in management of the grove, or additional costs for purchase or adaptation of machinery. This latter point makes the coppiced system suitable for areas of traditional culture characterized by old trees, obsolete training systems or small farms. This method can also be used to rejuvenate old trees, thus increasing productivity of obsolete groves. As a result, managing the olive orchard by coppicing has the following advantages: (a) reduction of labour costs for pruning; (b) no requirement of skilled labour; (c) low-cost implementation; (d) attenuation of alternate bearing; e) increased productivity of old trees; and (f) control of tree size, which facilitates management of the tree from the ground. The coppiced wood may represent an additional source of income in some areas.

The main disadvantage appears to be the multiple trunks, requiring several attachments for mechanical harvest. At present the coppiced orchard can only be recommended if plants are harvested by hand. The method of coppicing has not been tested on table cultivars, but it is likely that it will prove unsuitable. The ten-year periodicity for coppicing should not be considered fixed and should be adapted to different pedoclimatic conditions. In the experiment described, the optimal duration of the coppiced cycle was estimated at 8–9 years; in more fertile soils or under irrigated conditions the duration of the cycle might be extended to over 10 years.

8.8 Others

Y-trellis. The Y-trellis was developed in Italy in the late 1950s as a system suitable for very high-density plantings. The Y-trellis consists of a single-trunk tree with two primary branches inserted at 0.7–1.0 m in height on the trunk. Trees are planted at narrow spacings; the primary branches can be oriented in the direction of the row or perpendicular to the row (Fig. 8.18). Heading of the main axis is usually done during the first growing season to remove the central leader and stimulate growth of shoots that will form the primary branches. The main branches are inclined to about 45° by tying them to trainer sticks. The vigour of the shoots growing below the insertion point of primary scaffolds may have to be controlled by bending to avoid these shoots prevailing over the selected primary

Figure 8.18 Young tree trained to Y-trellis. The two main scaffold branches were obtained by cutting rather than inclination with trainers to reduce pruning costs.

branches. A few secondary branches are selected on each primary branch. The pruning of mature trees will consist mainly in renewing the fruiting shoots and secondary branches. Particular care must be taken to eliminate watersprouts from the upper proximal part of the inclined branch. Pruning the Y-trellis system is expensive, and plants are not more productive than if trained to other systems.

Fuse. The fuse is a single-trunk training system similar to the central leader, but with the lowermost branches the same length as the uppermost ones. The central leader arrow of the tree has to be periodically shortened by pruning to contain tree height. Because of the limited volume in the lower part of the canopy it has been recommended for very high-density plantings. However, there are no specific advantages in training trees to this shape with the currently available cultivars. The fuse will be more useful once compact cultivars or low-vigour rootstocks are available.

Hedge. The hedge is a system obtained by planting trees at narrow spacings (1.5 to 4 m) along the row to form a continuous fruiting wall. The canopy is allowed to grow freely during the training phase. Because of the short distance on the row, trees grow high and expand towards the inter-row. Pruning during the mature stage is limited to the removal of exhausted branches and opening gaps in the thick canopy. Mechanical pruning can be effectively used in a rotation with manual pruning. The hedge system is no more productive than other systems and it is not suitable for mechanical harvesting by shakers. In some areas it is used to form wind-breaks to protect crops more sensitive to salty aerosols than olive.

Chapter 9

Comparisons of training systems

9.1 General

In this chapter the results from the few published works that have compared the performance of different training systems are summarized. Most of these studies are recent and data refer to the first ten years after planting only. However, the results are quite clear and, in most cases, conclusive. Since not all systems have been properly tested so far, where published information was not available we reported comments and opinions based on our personal experience. Hopefully, future research will continue to collect data from the trials already under study and expand the experimental basis of comparison.

There is an almost complete lack of data with regard to the effect of the training system on physiological parameters in olive. The training system has been shown to affect light interception, leaf area, gas exchange, dry matter accumulation, and yield in apple and peach trees (Caruso et al., 1999; Giuliani et al., 1999; Jackson, 1980; Robinson and Lakso, 1991), but no studies have been published for olive. In a preliminary investigation, light interception and plant area index were estimated by fisheye photography in an orchard where trees planted at 6 x 6 m had been trained to either central leader, free vase, or bush. There were no differences in the plant area index and light interception on a hectare basis between the different systems (Gucci R., Cantini C., van Gardingen P. and Sharp L., unpublished results).

9.2 Technical comparison

There are a few studies published recently comparing the central leader and vase systems. In most cases more than one cultivar was tested and several parameters were measured, including vegetative growth, flowering, fruit yield, fruit size, oil yield, yield to mechanical harvesting, and onset of fruit production.

Proietti et al. (1998) studied the effect of the training system on growth, yield and oil quality in a drip-irrigated orchard planted at a 5 x 5 m distance. They showed that there were no differences in fruit yield (accumulated over eight years) between trees of 'Frantoio' and 'Leccino' trained to either central leader or vase, but trees of 'Maurino' trained to central leader yielded 40% less than those trained to a vase. This difference was probably due to the lesser increment in trunk cross-sectional diameter (TCSA) and the tendency of 'Maurino' to produce long thin shoots and dense canopies. No differences in TCSA due to the training system were found in 'Frantoio' and 'Leccino'. Oil quality (acidity, peroxide number, chlorophyll fatty acid composition, polyphenols content) was unaffected by the training system (Proietti et al., 1998).

Another comparison was conducted in an irrigated orchard in central Italy, where trees of three cultivars (Frantoio, Moraiolo, Nostrale di Rigali) had been planted at a 7 x 6 m distance and trained to either a central leader or a vase (Preziosi et al., 1994). There was no effect of the training system on shoot length, flowering (number of blossoms per fertile node, flowers per blossom, flowers per fruiting shoot) and oil characteristics (acidity, peroxide number, polyphenols content, stability to oxidation) in all three cultivars. Yield accumulated over the first three years of production was significantly higher in the vase of 'Frantoio' only. There was no effect of the training system on yield efficiency in 'Frantoio' and 'Moraiolo', but the central leader of 'Nostrale di Rigali' had a higher efficiency than the vase. Crown volume and trunk diameter were greater in plants trained to vase than in those trained to central leader of 'Moraiolo' and 'Frantoio'. More wood was removed by pruning the vase than the central leader trees (Preziosi et al., 1994).

Parlati et al. (1996) compared three training systems in a 5 x 5 m trial of three cultivars under dryland cultivation. They reported that trees trained to a vase produced more than central leader trees of 'Leccino', but no differences were found between these two systems in 'Frantoio' and 'Moraiolo'. On the other hand, the vase of 'Frantoio' was more productive than the Y-trellis system. Interestingly, there was no effect of the training system on flowering, fruit set, and fruit size, or on fatty acid composition, polyphenols content, acidity, and peroxide number of the oil. The yield efficiency of the canopy was higher in central leader trees than in both other systems (1.15 vs. 0.48 kg fruit m^{-3} canopy volume). The Y-trellis system did not show any advantage and was less productive than other systems (Parlati et al., 1996).

An extensive investigation on productivity and costs for cultural practices in modern olive orchards was promoted by the Regional Agency for Agricultural Development and Innovation of Tuscany (ARSIA) following the great frost of 1985. More than 25 orchards were newly planted at different sites with plants trained to either central leader or free-canopy systems (bush, STFC). In three sites where the performance of the bush was compared with that of the central leader within the same farm, the free-canopy system started production one year earlier and yielded more than central leader trees (Table 9.1). The bush also proved more efficient and cheaper than the central leader (see also section 9.3). These results were confirmed by considering the data from all the ARSIA plantings in Tuscany and the work by Silvestri et al. (1999). Only in one case did the onset of production occur later than the fourth year after planting in free-canopy systems, whereas it happened in 5 out of the 13 cases for central leader orchards (Tables 8.1 and 8.2). Under non-irrigated conditions at 6 x 6 m planting distance, the bush produced an average of 2.21 t/ha (fruit), whereas the central leader yielded an average of 1.57 t/ha (Tables 8.1; 8.2; 9.1). Under irrigated conditions central leader trees planted at 6 x 3 m distance yielded an average of 3.22 t/ha; in irrigated orchards where plants had been grown as free bush at 6 x 5 m distance an average yield of 2.75 t/ha was obtained (Tables 8.1; 8.2). It should be noted that the low yields obtained in some of these trials were due to the occurrence of frosts and the relatively short growing season of many sites.

Table 9.1 Comparison between central leader and bush systems planted at 6 x 6 m distance under non-irrigated conditions at three sites in Tuscany. Cultivars were 'Frantoio', 'Leccino' and 'Pendolino' at Cetona and Pomarance; 'Pendolino' was replaced by 'Moraiolo' at Suvereto. Central leader trees were pruned annually, the free-canopy plants only when needed. Onset of production was estimated when fruit yield exceeded 0.15 kg per plant. Source of raw data: ARSIA, 1999.

Site	Training system	Onset of production (years after planting)	Yield (Kg fruit/plant)	(t/ha)
Cetona	Bush	4	6.7	1.87
	Central leader	5	3.7	1.03
Pomarance^	Bush	2	8.2	2.27
	Central leader	4	1.6	0.44
Suvereto	Bush	3	10.4	2.88
	Central leader	4	7.4	2.05
Mean of trials	Bush	N.A.	8.4	2.34
	Central leader	N.A.	4.2	1.17

^ Low production due to frost damage in the whole orchard;
N.A. Not applicable.

Cantini and Sillari (1998b) compared the effect of the training system on production, alternate bearing, and oil quality under non-irrigated conditions. Although their investigation was conducted in different orchards it was clear that yield was not decreased by using free-canopy systems or reducing the frequency of pruning to biennial or 10-year cycles. Trees trained to polyconic vase or vasebush and pruned yearly were more expensive than plants grown as bush, or pruned every other year (Cantini and Sillari, 1998b). Morettini (1972) showed that trees trained to palmette started production later and produced less than plants grown as vasebush.

There are not many studies comparing yields from mechanical harvesting of different training systems. Fontanazza (1993) considered the central leader the most suitable system for mechanical harvesting by shakers, but he did not compare the central leader with other modern training systems. Famiani et al. (1998) showed that yield by mechanical shakers was approximately the same in trees trained to either free vase or central leader, which is in agreement with our experience. Problems inherent to the olive plant represent a limit for yield efficiency to mechanical harvesting regardless of the training system (see also section 7.2).

It is quite evident that, strictly from the technical point of view, excellent results can be obtained using many different training systems. Therefore, the best training system should allow the grower to maximise revenue, which, in turn, depends on reducing costs for pruning and harvesting. A valid alternative is offered by mechanical pruning, provided that it is correctly used in alternation with manual pruning or no pruning. On the other hand, mechanical pruning results in low productivity and an excessive density of foliage if used as the only pruning method over a few years (see section 6.6).

9.3 Economic comparison

The cost of pruning for the different training systems largely depends on the hours of labour required. Cantini and Sillari (1998b) summarized the number of hours required for different training systems using the ARSIA (1994) trials as the data source. The number of hours required to form and maintain central leader trees ranged from 95 to 925 hours over 9 years. Further analysis of these data updated to the 1998 growing season indicated that the time required for pruning trees to central leader ranged from 15 to 113 h of labour per hectare per year starting from the first year after planting (Table 8.1). On the other hand, the annual pruning time in the first 10–12 years after planting ranged from 3.5 to 21 hours of labour for plants grown as free-canopy systems. The high variability of the data was due to the ample sampling basis used in the original investigation aimed at including farms representative of the different cultural conditions in Tuscany. Thus, farm management, soil type and climatic conditions were not homogeneous. Nevertheless, the time needed to prune central leader trees was more than that required to form a bush across all trials. The average annual pruning time calculated for all the ARSIA trials (12 trained to central leader and 10 to free canopy) was 47 and 13 hours per hectare, respectively.

The economic performance of central leader, free-canopy systems and vase has been compared only in a few studies (Angeli et al., 1995; Cantini and Sillari, 1998a). At the current cost of oil and labour in Italy, it takes 10 years to break even for an orchard trained to a bush and 24 years for the central leader. The longer period for the investment return calculated for central leader is due to the higher pruning costs than for free-canopy systems. In addition, the high density plantings of 555 trees per hectare trained to a central leader proved uneconomical (Table 8.1; Cantini and Sillari, 1998b). It goes without saying that these estimates vary with economic conditions. The vasebush and free vase also are relatively economical in terms of pruning. The cheapest system has proved to be the coppiced but the experimental evidence so far documented needs to be tested further under other climatic conditions.

Similarly, further testing of mechanical pruning strategies needs to be conducted in the future. Mechanical pruning has proved very promising to reduce pruning costs. Giametta and Zimbalatti (1997) reported that pruning olive orchards can be sustainably maintained by 21 hours of manual pruning and 4 hours of mechanical pruning per hectare. More optimistic figures were reported by Fontanazza et al. (1998), who indicated that only 1.5 h per tree over 9 years were needed when mechanical pruning was alternated with no pruning and manual pruning in a triennial rotation.

In Table 9.2 we have estimated the time needed to prune plants to different training systems from the first year after planting. These values were derived from our experience and refer to a pruner of medium skill and experience. They should be considered merely indicative, since conditions may vary from site to site and from person to person. The longer time needed for pruning mature central leader trees is due mainly to the need to contain height by removing the original leader and replace it with a subsidiary leader. This operation is time-consuming because it often requires a ladder or a platform driven by tractor. We considered that removal of suckers took the same time for all systems.

Table 9.2 Time for hand-pruning olive plants trained to either free vase, central leader, single-trunk free-canopy, or bush. Times are given in minutes per plant.

Year	Free vase	Central leader	Single-trunk free-canopy	Bush
1	0	3	2	0
2–4	8	5	5	0
5–8	10	15	10	5
8–on	20^	20	15	15^

^ If mature trees are pruned every other year, this value is halved.

The bush, the vasebush or the coppiced management are only suitable for hand-harvesting, unless harvesting machines other than shakers are used. In many situations the high cost of hand-picking requires that fruits be harvested mechanically. The free vase, globe, central leader, or single-trunk free-canopy are the most suitable training systems for harvesting by shakers. Famiani et al. (1998) estimated that mechanical harvesting of central leader trees took about 10% longer than trees trained to a vase, because the low-hanging shoots of the central leader made attachment of the shaker's clamp more difficult. Hence, these shoots should be removed by pruning both in central leader and free-canopy systems with a single trunk, not to slow down the work of shakers (Fig. 8.11a; 8.13b). Single-trunk free-canopy systems, which combine the concepts of minimum pruning with the requirement of the single trunk for the shaker's attachment, are formed and maintained economically.

What's the best training system?

This often-asked question does not have a single answer. We have already remarked in Chapters 7, 8 and 9 that the choice of the training system depends on the characteristics and objectives of the orchard. It is impossible to select the training system without considering such factors as the method of harvesting, planting density, and product use. However, a few general comments can be made here.

First of all, since many training systems allow the production of high yields of excellent quality, the best system for olive growing is the one that can be obtained at minimum cost without negative side effects on plant performance and orchard management. Mechanical pruning is a valid alternative, provided it is used in combination with manual pruning and no pruning. Mechanical pruning can be adapted to vase, globe, central leader, and free-canopy systems.

If plants are to be harvested manually for oil production, the pruning should be as simple and rapid as possible. The best training systems are the vasebush, the free vase, and the bush, all compatible with minimum pruning strategies. All of these systems allow management of the tree almost entirely from the ground. The free vase and the vasebush require more pruning than the bush during the training period. However, we have shown in sections 6.3 and 8.6 that both the vasebush and the free vase can be formed starting from a canopy which has received little pruning or no pruning during the first years after planting. The bush allows maximum flexibility in pruning at low costs. Once plants grown as free-canopy systems become fully productive, biennial pruning can be adopted and sustained over many years. Studies in progress have shown that pruning every other year does not result in lower yields or quality over 15 years (Cantini and Sillari, 1998b). It is likely that this holds true over the entire economic life of the orchard, but as yet there are no data documenting this.

Maximum reduction in pruning costs can be obtained by coppicing the orchard or pruning every two or three years; however, these methods should be used after adequate experience in olive growing. The coppicing method has not been tested for the production of table olives, but it is likely that it would not be suitable. Table olive production requires that access to the different sectors of the canopy be easy for harvesting. As a result, manual pruning will be more time-consuming than for oil cultivars and includes the summer pruning necessary to thin the crop load (section 6.4).

A single trunk is indispensable for harvesting trees by mechanical shakers. The vase and the central leader are equally suitable for mechanical harvesting and the main limit remains the high cost of pruning. Both training systems should be interpreted as freely as possible. It is quite evident that the detailed pruning required for the central leader or traditional vase systems is not justified by technical advantages or economic convenience. Single-trunk free-canopy systems (STFC) appear to be an excellent solution for combining the single trunk with minimum pruning costs. Although STFC systems are less demanding than other single-trunk systems in terms of the pruning prescribed, they still require experienced pruners.

In the literature there is no evidence that increasing the planting density to over 420 trees/ha under irrigated conditions improves the economic performance of the orchard. In most cases densities of 277 or 333 trees/ha are the upper limit beyond which costs increase more than revenues. Very high-density plantings (555 trees/ha or more) were originally proposed for an early onset of production and high yields (Fontanazza, 1993). However, when comparisons were made with other training systems or lower planting densities, the 6 x 3 m central leader orchards proved less economically convenient since the higher yields did not compensate for the greater initial investment in plants and pruning (Angeli et al., 1995; ARSIA, 1999).

Central leader systems have been promoted as the best solution for intensive mechanized orchards (Fontanazza, 1993), but all the published studies comparing central leader with other modern training systems (free vase, single-trunk free-canopy, bush) have shown that there was no evidence of higher productivity, growth, or yield efficiency to mechanical harvesting (Tables 8.1 and 8.2; ARSIA, 1999; Cantini and Sillari, 1998b; Parlati et al., 1996; Preziosi et al., 1994; Proietti et al., 1998; Tombesi, 1996). In a few cases central leader proved even less productive than other training systems (Preziosi et al., 1994; Proietti et al., 1998; Tombesi, 1996).We believe that an important factor in the success of the central leader in recent years lies in the more regular architecture of the canopy and orderly appearance of trees—an aesthetic advantage over free-canopy systems. It is interesting that almost 30 years ago Morettini (1972) wrote that "the central leader should not be recommended for the olive because, as years go by, one obtains an elegant shape but plants too high to be easily harvested by hand or machine".

Chapter 10

Conclusions

The olive orchard is a long-term investment. In the past the economic life of the olive orchard coincided with the longevity of the tree and exceptional events such as frosts or droughts determined the pace of crop renovation. Today, the economic life of the orchard is no more than 40 years. When visiting traditional orchards, it is quite common to see a variety of tree shapes and sizes. The evolution of training systems has been slow over the ages also because of socio-economical factors (see also Chapter 1), and we focused only on those economically feasible today. Several valid alternatives are available. For hand-picking, the best training systems are the bush, the vasebush, and the vase. The coppiced management system represents an extreme case of how the olive canopy and the whole orchard can be managed.

A high degree of mechanization of harvesting is important, especially in large plantations. Considerable improvements have been made in the design and development of new mechanical harvesters, but to improve efficiency of shaking and fruit recollection it is necessary that trees be trained to a single trunk. The vase, free vase, globe, and central leader systems are equally suitable for mechanical harvesting and the main constraint remains the high pruning cost. In this respect, good results can be obtained with STFC, which produces abundantly and yields to mechanical harvesting similarly to other systems, but at a reduced cost of pruning. It is quite evident that, strictly speaking from the technical point of view, excellent results can be obtained using many different training systems.

In recent years there has also been a tendency towards very high-density plantings of more than 500 trees per hectare to maximize yields. However, the economic results showed that costs increased more than incomes under both irrigated and non-irrigated conditions so that this level of cultural intensification is unjustified economically. Very high-density plantings can become profitable only when new cultivars or clones with a compact growth habit and high harvest index are developed by breeding or clonal selection.

Despite improvements in machinery for mechanical harvesting, manual harvesting is still the most common method to harvest olive fruits in many olive-growing areas. In Italy it has been estimated that only 2% of the olive orchards are harvested mechanically and less than 10% by hand-operated mechanical combs. Sociological and organizational conditions (small size of orchards, family-owned groves), environmental and physiological constraints to mechanization (steep slopes, old groves), and low cost of labour in poorer areas are some of the main reasons for the slow uptake of mechanical harvesting. Nevertheless, hand-picking is expensive and can be as much as 80% of the annual cost of cultivation. Because of the high cost of manual harvesting, it will be just a matter of time before the use of harvesting machines becomes more widespread in the olive orchard. This process will be faster in countries where the size of orchards is large.

Although at the moment shakers represent the best technology for harvesting, they are far from being the optimal solution. The two main constraints are the low yield efficiency and the long time required to harvest each tree. Fruit interception on nets is very time-consuming, whereas shakers with catching frames require wide spacings between trees to work efficiently. Great improvements in efficiency and a reduction in the time for harvesting can be obtained with machines harvesting continuously by combing/whipping the canopy surface from the top or along the side of the row. With regard to devices for the recollection of fruits once they have been detached, the prototypes developed so far still need improvement. But the greatest limitation to the development of these new machines is the inadequacy of the current olive cultivars, which are too large and vigorous to be maintained within volumes compatible with the size of harvesting machines. Once compact cultivars and rootstocks become available (they are currently unavailable), it is likely that the continuous mechanical harvesting along the row will become the most convenient method for olive orchards. In that case the orchard design will have to be modified to using training systems that reduce the expansion of canopy volume and high planting densities for a uniform fruiting surface.

In modern olive-growing, inputs and resources should be optimized for profitable cultivation in the long run. It should be apparent by now that there are several alternatives for pruning or training trees. Solutions should be devised at the farm level or at a small territorial scale, where cultural conditions are fairly homogeneous. Solutions valid for all cultivars, climates, soils and social conditions are not realistic. The grower should decide on the pruning strategy to achieve abundant yields and a high quality of production, but should consider the optimization of resources and economic viability of the adopted techniques and methods as well.

Olive growing is evolving fast and so is the market for its products. Changes in orchard management have occurred and are occurring as a result of this evolution. Nowadays, plantings are denser on a land unit basis, irrigation is more widely used than in the past, and mechanization of harvest and cultural practices have reached satisfactory levels. Since harvesting and pruning are the two most expensive practices in the olive orchard, pruning techniques and training systems are continually being re-evaluated. Recent developments include the adoption of single-trunk systems for mechanical harvesting and free-canopy systems to reduce pruning costs. Mechanical pruning also appears very promising as an additional way to keep pruning costs low, provided that it is correctly integrated with manual pruning or no pruning.

Today, there are many types of olive growing. The small olive farm, typical of most olive-growing areas around the Mediterranean basin, is not going to disappear as long as there is a market for high-quality products—both table olives and oil. Family-operated olive farms will probably not be restricted to the traditional olive-growing countries, but will also be viable in some of the newly developing areas. On the other hand, most of the olives in these "new" countries will be produced by large enterprises competing for market share and operating on a completely different scale from that of the family-owned farm. Olive growing on a large scale requires adjustments in cultural management and technological innovation. No matter how traditional olive growing may still appear to be in many areas, this industry is very different today from what it was for centuries.

References

Alegre S., Girona J., Marsal J., Arbones A., Mata M., Montagut D.,Teixido F., Motilva M.J., Romero M.P. 1999. Regulated deficit irrigation in olive trees. *Acta Horticulturae* **474** (1): 373–6.

Angeli L., Sillari B., Cantini C. 1995. Cespuglio e monocono a confronto. *Informatore Agrario* **51** (43): 59–63.

Angelopoulos K., Dichio B., Xiloyannis C. 1996. Inhibition of photosynthesis in olive trees (*Olea europaea*) during water stress and rewatering. *Journal of Experimental Botany* **47**: 1093–100.

Antognozzi E., Cartechini A., Tombesi A., Proietti P. 1990. Effect of cultivar and vibrator characteristics on mechanical harvesting of olives. *Acta Horticulturae* **286**: 417–20.

ARSIA. 1994. *Prove sperimentali e dimostrative—Olivicoltura*. Regione Toscana, Firenze.

ARSIA. 1999. *Dieci anni di sperimentazione olivicola in Toscana*. Effemme Lito. Firenze. 167 pp.

Bardi G. 1802. *Sulla più vantaggiosa forma da darsi nella potatura agli ulivi*. Archivio Storico Accademia Georgofili no. 269. Firenze. 18 pp.

Barranco D., Milona G., Rallo L. 1994. Epocas de floracion de cultivares de olivo en Còrdoba. *Investigacion Agraria* **9** (2): 213–20.

Barranco D., Fernandez-Escobar R., Rallo L. 1997. *El cultivo del olivo*. 2nd edition. Mundi-Prensa, Madrid. 651 pp.

Bartolini G., Prevost G., Messeri C., Carignani G. 1998. Olive germplasm. Cultivars and world-wide collections. FAO, Rome. 459 pp.

*Battarra G. 1782. *Pratica agraria*, 2nd Edition. Cesena, Italy. (Reprinted in 1975 by Chigi, Rimini.)

Ben-Tal Y., Lavee S. 1984. Girdling olive trees, a partial solution to biennial bearing. II. The influence of consecutive mechanical girdling, on flowering and yield. *Rivista Ortoflorofrutticoltura Italiana* **68**: 441–51.

Bongi G., Palliotti A. 1994. Olive. In *Handbook of environmental physiology of fruit crops* (B. Schaffer and P.C. Andersen, eds). CRC Press, Boca Raton FL, USA. pp. 165–87.

Bongi G., Mencuccini M., Fontanazza G. 1987. Photosynthesis of olive leaves: effects of light flux density, leaf age, temperature, peltates, and H_2O vapour pressure deficit on gas exchange. *Journal of the American Society for Horticultural Science* **112**: 143–8.

Breviglieri N. 1961. La nuova olivicoltura specializzata intensiva. *Italia Agricola* **98** (3): 215–69.

Brighigna A. 1998. *Le olive da tavola*. Edagricole, Bologna. 205 pp.

Cantini C., Gucci R., Sillari B. 1998. An alternative method to managing olive orchards: the coppiced system. *HortTechnology* **8** (3): 409–12.

Cantini C., Sillari, B. 1998a. Risultati produttivi ed economici di oliveti condotti con diversi sistemi di potatura. *Rivista Frutticoltura* **60** (1): 49–54.

Cantini C., Sillari B. 1998b. Esperienze toscane nell'intensificazione colturale dell'olivo. In *Seminari di olivicoltura*, Accademia Nazionale dell'Olivo. Spoleto, Italy. pp. 127–39.

Caruso T., Inglese P., Sottile F., Marra F.P. 1999. Effect of planting system on productivity, dry matter partitioning and carbohydrate content in above-ground components of 'Flordaprince' peach trees. *Journal of the American Society for Horticultural Science* **124**: 39–45.

Cimato A. 1989. Riflessi dei fattori agronomici sulle caratteristiche qualitative delle olive da tavola. *Rivista Frutticoltura* **51** (11): 17–22.

Civantos L., M. Pastor. 1996. Production techniques. In *World olive encyclopaedia*. International Olive Oil Council, Madrid. pp. 147–94.

Colorio G. 1997. Raccolta meccanica delle olive. *Informatore Agrario* (44): 43–9.

Crider F.J. 1922. The olive in Arizona. *The University of Arizona, Agricultural Experiment Station, Bulletin* no. 94, Tucson AZ, USA. pp. 491–528.

*Davanzati B. 1807. *Coltivazione toscana delle viti e d'alcuni arbori*. Società Classici Italiani, Milano. 89 pp. (Reprinted in 1978 by Fogola, Torino.)

Denney J.O., Martin G.C., Kammereck R., Ketchie D.O., Connell J.H., Krueger W.H., Osgood J.W., Sibbett G.S., Nour G.A. 1993. Freeze damage and cold hardiness in olive: findings from the 1990 freeze. *California Agriculture* **47** (1): 1–12.

Famiani F., Proietti P., Palliotti A., Guelfi P., Nottiani G. 1998. Possibilità di meccanizzazione della raccolta delle olive in diverse tipologie di oliveto. *Rivista Frutticoltura* **60** (7–8): 33–9.

Faust M. 1989. *Physiology of temperate zone fruit trees*. John Wiley & Sons, New York, 338 pp.

Ferguson L., Reyes H., Metheney P. 1999. Mechanical harvesting and hedging of California black ripe (*Olea europaea*) cv. 'Manzanillo' table olives. *Acta Horticulturae* **474** (1): 193–6.

Ferguson L., Sibbett G.S., Martin G.C. 1994. *Olive production manual*. University of California, Division of Agriculture and Natural Resources, Oakland CA. Publication 3353. 156 pp.

Fernàndez-Escobar R., Benlloch M., Navarro C., Martin G.C. 1992. The time of floral induction in the olive. *Journal of the American Society for Horticultural Science* **110**: 303–9.

Fontanazza G. 1993. *Olivicoltura intensiva meccanizzata*. Edagricole, Bologna. 312 pp.

Fontanazza G., Camerini F., Bartolozzi F. 1998. Intervento meccanico e manuale nella potatura di produzione. *Olivo e Olio* **1**: 27–34.

Francolini F. 1935. La ricostituzione dell'olivo. *Italia Agricola* **72** (12): 993–1006.

Garcia-Ortiz 1998. Poda del olivo. In *L'olivicoltura mediterranea verso il 2000—Atti del VII International Course on Olive Growing* (A. Cimato, A. Baldini, eds). Tip. A-emme, Scandicci (FI). pp. 131–45.

Garcia-Ortiz, A., Fernandez A., Pastor M., Humanes J. 1997. Poda. In *El cultivo del olivo*. 2nd edition. (Barranco D., Fernandez-Escobar R., Rallo L., eds). Mundi-Prensa, Madrid. pp. 307–43.

Garrido Fernandez A., Fernandez Diez M.J., Adams M.R. 1997. *Table olives: production and processing*. Chapman & Hall, London. 495 pp.

Giametta G., Zimbalatti G. 1994. Possibilities of mechanical pruning in traditional olive groves. *Acta Horticulturae* **356**: 311–14.

Giametta G., Zimbalatti G. 1996. La meccanizzazione della potatura in olivicoltura. In *Atti del convegno "L'olivicoltura Mediterranea: stato e prospettive della coltura e della ricerca"*. La Grafica Commerciale, Cosenza, Italy. pp. 365–73

Giametta G., Zimbalatti G. 1997. Mechanical pruning in new olive-groves. *Journal of Agricultural Engineering Research* **68**: 15–20.

Gifford R.M., Thorne J.H., Hitz W.D., Giaquinta R.T. 1984. Crop productivity and photoassimilate partitioning. *Science* **225**: 801–7.

Girona J. 1995. *Requerimientos hidricos del olivo. Estrategias de aplicacion de cantidades limitadas de aqua de riego en 'Arbequina'*. Les Borges Blanques (Lleida). Ponencias y Comunicaciones 67–71.

Giuliani R., Magnanini E., Corelli Grappadelli L. 1999. Relazioni tra scambi gassosi e intercettazione luminosa in chiome di pesco allevate secondo tre forme. *Rivista Frutticoltura* **61** (3): 65–9.

Goldhamer D.A., Dunai J., Ferguson L.F. 1994. Irrigation requirements of olive trees and responses to sustained deficit irrigation. *Acta Horticulturae* **356**: 172–5.

Goldhamer D.A. 1999. Regulated deficit irrigation for California canning olives. *Acta Horticulturae* **474** (1): 369–71

Grammatikopoulos G., Karabourniotis G., Kyparissis A., Petropoulou Y., Manetas Y. 1994. Leaf hairs of olive (*Olea europaea*) prevent stomatal closure by ultraviolet-B radiation. *Australian Journal of Plant Physiology* **21**: 293–301.

Gucci R. 1998a. Assimilazione e ripartizione del carbonio in foglie di olivo. *Rivista Frutticoltura* **60** (7/8): 77–82.

Gucci R. 1998b. Effetto dei fattori ambientali sulla produttività dell'olivo. In *L'olivicoltura mediterranea verso il 2000—Atti del VII International Course on Olive Growing* (A. Cimato, A. Baldini, eds). Tip. A-emme, Scandicci (FI). pp. 207–14.

Gucci R., Cantini C., van Gardingen P., Sharp L. 1999. Determination of the plant area index of olive trees by hemispherical photography. *Acta Horticulturae* **374** (1): 317–19.

Gucci R., Tattini M. 1997. Salinity tolerance in olive leaves. *Horticultural Reviews* **21**: 177–214.

Guerriero R., Scaramuzzi F., Crescimanno F. G., Sottile I. 1972. Ricerche comparative tra olivi innestati ed autoradicati. Osservazioni nei primi anni dall'impianto. *Tecnica Agricola* **24** (4): 3–27.

Guerriero R., Vitagliano C. 1973. Influenza dell'ombreggiamento sulla fruttificazione dell'olivo. *Agricoltura Italiana* **73**: 85–115.

Harris R.W. 1994. Clarifying certain pruning terminology: thinning, heading, pollarding. *Journal of Arboriculture* **20**: 50–4.

Hartmann H.T., Opitz K.V., Bentel J.A. 1960. La taille des oliviers en California. *Informations Oléicoles Internationales* **11**: 33–67.

Iannotta N., Lombardo N., Briccoli Bati C., Monardo D. 1999. Risposta di alcune cultivar di olivo alle minime termiche tardive. *Informatore Agrario* **55** (14): 59–62.

International Olive Oil Council (IOOC). 1996. *World encyclopaedia of the olive*. Madrid. 479 pp.

Jackson J.E. 1980. Light interception and utilization by orchard systems. *Horticultural Reviews* **2**: 208–67.

Kramer P.J, Kozlowski T.T. 1979. *Physiology of woody plants*. Academic Press, London. 811 pp.

Lambers A.F., Chapin F.S., Pons T.L. 1998. *Plant physiological ecology*. Springer Verlag, New York, 540 pp.

*Landeschi G.B. 1775. *Saggi di Agricoltura*. Cambiagi, Firenze. 293 pp. (Reprinted in 1998 by ETS, Pisa.)

Lang G.A., Early J.D., Martin G.C., Darnell R.L. 1987. Endo-, para-, and ecodormancy: physiological terminology and classification for dormancy research. *HortScience* **22** (3): 371–7.

Lavee S. 1990. Aims, methods and advances in breeding of new olive (*Olea europaea* L.) cultivars. *Acta Horticulturae* **286**: 23–36.

Lavee S. 1986. Olive. In *Handbook of fruit set and development* (R. Monselise, ed.), CRC Press, Boca Raton FL, USA. pp. 261–76.

Lavee S. 1994. The Israeli olive industry: present state, problems and research activities. In *Atti del convegno "L'olivicoltura Mediterranea: stato e prospettive della coltura e della ricerca"*. La Grafica Commerciale, Cosenza, Italy. pp. 59–76.

Lavee S. 1996. Biologia e fisiologia dell'olivo. In *Enciclopedia mondiale dell'olivo*. International Olive Oil Council, Plaza & Janés Editores, Barcelona. pp. 59–110.

Lavee S., Haskal A., Ben-Tal Y. 1983. Girdling olive trees, a partial solution to biennial bearing. I. Methods, timing and direct tree response. *Journal of Horticultural Science* **58** (2): 209–18.

Lo Gullo M.A., Salleo S. 1988. Different strategies of drought resistance in three Mediterranean sclerophyllous trees growing in the same environmental conditions. *New Phytologist* **108**: 267–76.

Lo Gullo M.A., Salleo S. 1990. Wood anatomy of some trees with diffuse- and ring-porous wood: some functional and ecological interpretations. *Giornale Botanico Italiano* **124**: 601–13.

Lòpez-Rivares E.P., Suarez M.P. 1990. Estudio de las epocas y anchuras optimas de anillado en olivo. *Olivae* **32**: 38–41.

Loussert R., Brousse G. 1978. *L'olivier*. Maisonneuve & Larose, Paris. 465 pp.

Martin G.C. 1987. Apical dominance. *HortScience* **22** (5): 824–33.

Martin G.C., Ferguson L., Polito V.S. 1994a. Flowering, pollination, fruiting, alternate bearing, and abscission. In *Olive production manual* (L. Ferguson, G.S. Sibbett, G.C. Martin, eds). University of California, Division of Agriculture and Natural Resources, Oakland CA. Publication 3353. pp. 51–6.

Martin G.C., Klonski K., Ferguson L. 1994b. The olive harvest. In *Olive production manual* (L. Ferguson, G.S. Sibbett, G.C. Martin, eds). University of California, Division of Agriculture and Natural Resources, Oakland CA. Publication 3353. pp. 117–28.

Meriño J. 1987. The costs of growing and maintaining leaves of mediterranean plants. In *Plant response to stress—functional analysis in Mediterranean ecosystems* (J.D. Tenhunen, F.M. Catarino, O.L. Lange, W.C. Oechel, eds). Nato ASI Series, Springer-Verlag, Heidelberg. pp. 553–64.

Mladar N., Kovacevic I. 1990. Experiment on regeneration of olive orchards in Hvar Island. *Acta Horticulturae* **286**: 275–7.

Monselise S.P., Goldschmidt E.E. 1982. Alternate bearing in fruit trees. *Horticultural Reviews* **4**: 128–73.

Morettini 1972. *Olivicoltura*. 2nd edition. REDA, Rome. 522 pp.

Orgaz F., Fereres E. 1997. Riego. In *El cultivo del olivo*. 2nd edition. (Barranco D., Fernandez-Escobar R., Rallo L., eds). Mundi-Prensa, Madrid. pp. 259–80.

O'Rourke M.K., Buchmann S.L. 1986. Pollen yield from olive trees cvs. Manzanillo and Swan Hill in closed urban environments. *Journal of the American Society for Horticultural Science* **111**: 980–4.

Parlati M.V., Iannotta N., Pandolfi S. 1996. Studio dell'influenza della forma di allevamento sul comportamento vegetativo e produttivo dell'olivo. In *Atti del convegno "L'olivicoltura Mediterranea: stato e prospettive della coltura e della ricerca"*. La Grafica Commerciale, Cosenza, Italy. pp. 343–53.

Pastor M. 1989. *La taille de l'olivier. Manuel Pratique d'Oleiculture*. International Olive Oil Council (IOOC), Madrid, 111 pp.

Pinney K., Polito V.S. 1990. Flower initiation in 'Manzanillo' olive. *Acta Horticulturae* **286**: 203–5.

Pochi D., Limongelli R., Vannucci D. 1996. Potatura meccanica dell'olivo. *Informatore Agrario* **52** (44): 45–8.

*Presta G. 1794. Degli ulivi, delle ulive, e della maniera di cavar l'olio. In *Opere di Giovanni Presta*, vol. II (Cavallera H.A., ed.). (Reprinted in 1989 by Edizioni del Grifo, Lecce, Italy.)

Preziosi P., Proietti P., Famiani F., Alfei B. 1994. Comparisons between 'monocone' and 'vase' training systems on the olive cultivars 'Frantoio', 'Moraiolo' and 'Nostrale di Rigali'. *Acta Horticulturae* **356**: 306–10.

Proietti P., Palliotti A., Famiani F., Preziosi P., Antognozzi E. 1998. Confronto tra le forme di allevamento a monocono e a vaso in diverse cultivar d'olivo. *Rivista Frutticoltura* **60** (7–8): 69–72.

Rallo L. 1997. Fructification y produccion. In *El cultivo del olivo* (D. Barranco, R. Fernandez-Escobar, L. Rallo, eds). 2nd edition. Mundi-Prensa, Madrid. pp. 115–44.

Rallo L. 1999. Miglioramento delle risorse genetiche. In *Proceedings of the international seminar on scientific innovation and their application in olive growing and elaiotechnique.* 10–12 March 1999, Firenze. pp. 1–28.

Rallo L., Martin G.C. 1991. The role of chilling in releasing olive floral buds from dormancy. *Journal of the American Society for Horticultural Science* **116**: 1058–62.

Rallo L., Suarez M.P. 1989. Seasonal distribution of dry matter within the olive fruit-bearing limb. *Advances in Horticultural Science* **3**: 55–9.

Rallo L., Torreno P., Vargas A., Alvarado J. 1994. Dormancy and alternate bearing in olive. *Acta Horticulturae* **356**: 127–36.

Rapoport H.F. 1997. Botanica y morfologia. In *El cultivo del olivo* (D. Barranco, R. Fernandez-Escobar, L. Rallo, eds). 2nd edition. Mundi-Prensa, Madrid. pp. 35–60.

Rapoport H.F., Rallo L. 1991. Postanthesis flower and fruit abscission in 'Manzanillo' olive. *Journal of the American Society for Horticultural Science* **116**: 720–3.

Robinson T.L., Lakso A.N. 1991. Bases of yield and production efficiency in apple orchard systems. *Journal of the American Society for Horticultural Science* **116**: 188–94.

Roventini A. 1936. La ricostituzione olivicola attraverso la potatura. *Italia Agricola* **7**: 517–27.

Salisbury F.B., Ross C.W. 1978. *Plant Physiology.* Wadsworth Publishing Co., Belmont CA, USA. 422 pp.

Salleo S., Nardini A. 1999. Ecofisiologia di *Olea oleaster* Hoffmgg. et Link: verso un modello predittivo dell'adattamento all'aridità. *Olivo e Olio* **2** (4): 70–9.

Sillari B. 1966. Note di olivicoltura nel Follonichese e nella bassa di Massa Marittima. Tip. S. Anna, Massa Marittima (GR). 18 pp.

Silvestri E., Bazzanti N., Toma M., Cantini C. 1999. Effect of training system, irrigation and ground cover on olive crop performance. *Acta Horticulturae* **474** (1): 173–5.

Sismondi J.C.L. 1995. *Quadro dell'olivicoltura toscana* (translated from *Tableau de l'agriculture toscane.* 1801. J.J. Paschoud, Genève). ETS, Pisa. 250 pp.

Societe National d'Oleiculture de France. 1913. *L'olivier et l'huile d'olive.* Paris. 320 pp.

Soderini G. 1817. *Trattato degli arbori. Prima parte.* Stamperia del Giglio, Firenze. 251 pp.

Soule J. 1985. *Glossary for horticultural crops.* John Wiley & Sons, New York. 898 pp.

Sutter E.G. 1994. Olive cultivars and propagation. In *Olive production manual* (L. Ferguson, G.S. Sibbett, G.C. Martin, eds). University of California, Division of Agriculture and Natural Resources, Oakland CA. Publication 3353. pp. 23–9.

Tavanti G. 1819. *Trattato teorico-pratico completo sull'ulivo.* vol. 1. Stamperia Piatti, Firenze. 254 pp.

Tombesi A. 1988. Intercettazione luminosa ed efficienza produttiva dell'olivo. *Rivista Frutticoltura* **50** (3): 21–6.

Tombesi A. 1989. Potatura e forme di allevamento nell'olivo. *Rivista Frutticoltura* **51** (1): 7–14.

Tombesi A. 1996. La raccolta meccanica delle olive. *Rivista Frutticoltura* **58** (2): 31–5.

Tombesi A. 1998. Impianto dell'oliveto e forme di allevamento. In *L'olivicoltura mediterranea verso il 2000—Atti del VII International Course on Olive Growing* (A. Cimato, A. Baldini, eds). Tip. A-emme, Scandicci (FI). pp. 115–29.

Tombesi A., Cartechini A. 1986. L'effetto dell'ombreggiamento della chioma sulla differenziazione delle gemme a fiore dell'olivo. *Rivista Ortoflorofrutticoltura Italiana* **70**: 277–85

Tombesi A., Jacoboni N. 1974. Le forme di allevamento idonee alla meccanizzazione della raccolta delle olive. Atti Incontro Frutticolo sulla "Raccolta meccanica delle olive", 12 dicembre 1974, Foligno, Italy.

Tombesi A., Standardi A. 1977. Effetti dell'illuminazione sulla fruttificazione dell'olivo. *Rivista Ortoflorofrutticoltura Italiana* **61**: 368–80.

Tonini S. 1937. *Note pratiche per la razionale coltivazione dell'olivo*. Tipografia Perugina, Perugia, Italy. 141 pp.

Tous J., Romero A., Plana J., Baiges F. 1999. Planting density trial with 'Arbequiña' olive cultivar in Catalonia (Spain). *Acta Horticulturae* **474** (1): 177–80.

Vannucci D., Limongelli R. 1996. Raccolta delle olive con macchine agevolatrici. *Informatore Agrario* **52** (44): 51–4.

Vettori P. 1762. *Trattato delle lodi, e della coltivazione degli ulivi*. Stamperia Stecchi, Firenze.

Vitagliano C. 1969. Osservazioni su rami di olivi potati "a siepe". *Agricoltura Italiana* 7–8.

Vitagliano C., Viti R., Scalabrelli G. 1983. Osservazioni quinquennali su alcuni interventi di ristrutturazione dell'olivo per aumentare l'efficienza della raccolta meccanica. *Rivista Ortoflorofrutticoltura Italiana* **67**: 375–85.

Zucconi F. 1994. Fisiologia ed etologia della potatura. In *Metodi innovativi di allevamento dei fruttiferi a ridotta richiesta di manodopera* (F. Zucconi and D. Neri, eds). ERSO, Tip. Emme.Bi, Padova, pp. 19–36.

* Books reprinted in offset.

Index